T0139792

Yield-Aware Analog IC Design and Optimization
in Nanometer-scale Technologies

António Manuel Lourenço Canelas
Jorge Manuel Correia Guilherme
Nuno Cavaco Gomes Horta

Yield-Aware Analog IC Design and Optimization in Nanometer-scale Technologies

 Springer

António Manuel Lourenço Canelas
Instituto Superior Técnico
Instituto de Telecomunicações
Lisbon, Portugal

Jorge Manuel Correia Guilherme
Instituto Politécnico de Tomar
Instituto de Telecomunicações
Lisbon, Portugal

Nuno Cavaco Gomes Horta
Instituto Superior Técnico
Instituto de Telecomunicações
Lisbon, Portugal

ISBN 978-3-030-41538-9   ISBN 978-3-030-41536-5   (eBook)
https://doi.org/10.1007/978-3-030-41536-5

© Springer Nature Switzerland AG 2020
This work is subject to copyright. All rights are reserved by the Publisher, whether the whole or part of the material is concerned, specifically the rights of translation, reprinting, reuse of illustrations, recitation, broadcasting, reproduction on microfilms or in any other physical way, and transmission or information storage and retrieval, electronic adaptation, computer software, or by similar or dissimilar methodology now known or hereafter developed.
The use of general descriptive names, registered names, trademarks, service marks, etc. in this publication does not imply, even in the absence of a specific statement, that such names are exempt from the relevant protective laws and regulations and therefore free for general use.
The publisher, the authors, and the editors are safe to assume that the advice and information in this book are believed to be true and accurate at the date of publication. Neither the publisher nor the authors or the editors give a warranty, expressed or implied, with respect to the material contained herein or for any errors or omissions that may have been made. The publisher remains neutral with regard to jurisdictional claims in published maps and institutional affiliations.

This Springer imprint is published by the registered company Springer Nature Switzerland AG
The registered company address is: Gewerbestrasse 11, 6330 Cham, Switzerland

Preface

Developments over the last decades in very large-scale integration technologies allowed meeting the increasing demand for faster, cheaper, and reliable electronic devices. One of the key factors to support those developments is the implementation of most high-level functions of the chip in digital circuitry, whose design is highly automated due to the adoption of mature electronic design automation (EDA) tools. While digital integrated circuits (ICs) design is mostly automated, its analog counterpart is supported by a set of independent tools, dedicated to each step of the design flow, and highly dependent on human intervention. The increasing demand in circuit performances and complexity of device models due to the aggressive IC technology down-scaling have led to the acceptance of new simulation-based optimization tools for analog IC sizing, thus increasing analog IC design flow efficiency. Most of those tools consider only nominal circuit parameters values during the optimization process. As devices shrink down into nanometer scale, the effects of process variation have become very important and not considering those effects during the optimization and sizing process may result in circuit solutions push to limits of performances and dangerously close to the boundary of feasibility. Therefore, including a prediction of the percentage of circuits that comply with circuit specifications after fabrication, i.e., the circuit parametric yield, in the sizing and optimization process is an important step to avoid expensive redesign iterations. Monte Carlo (MC) analysis is the most general and reliable technique for yield estimation, yet the considerable amount of time it requires has discouraged its adoption in population-based analog IC circuit sizing and optimization tools.

The new yield estimation methodology developed and presented in this book is able to reduce the time impact caused by MC simulations in the context of analog ICs yield estimation, enabling its adoption in optimization processes with population-based algorithms, such as genetic algorithm (GA), considering the yield as one of the optimization problem objectives. The proposed methodology reduces the total number of MC simulations required to evaluate the optimization algorithm population. The reduction in the total number of simulations is achieved because at each GA generation the population is clustered and only the representative individual from each cluster is subject to MC simulations. Initial tests using a modified version of the k-means clustering algorithm, to identify similar individuals in the GA population, and a new technique to select the cluster representative individuals were able to achieve a reduction rate up to 91% in the total number of MC simulations, when compared to the number of MC simulations required to evaluate the complete GA population.

The need to balance the trade-off between yield estimation accuracy and the reduction rate of MC simulations made the k-means methodology to evolve and search for different clustering techniques. A new version of the developed yield estimation methodology with reduced time impact from MC simulations was finally developed and implemented in a state-of-the-art analog IC sizing tool using the fuzzy c-means clustering algorithm. The new methodology based on fuzzy c-means and named FUZYE is able to achieve good yield estimation accuracy and high reduction rates in MC simulations. The FUZYE methodology shows that the yield for the rest of the nonsimulated individuals in the population can be accurately estimated based on the membership degree of fuzzy c-means and the cluster representative individuals yield values alone. This new method was applied on several circuit sizing and optimization problems, and the obtained results were compared to the exhaustive approach, where all individuals of the population are subject to MC analysis. The FUZYE methodology presents on average a reduction of 89% in the total number of MC simulations, when compared to the exhaustive MC analysis over the full population. Moreover, other important clustering algorithms were also tested and compared with the proposed FUZYE, with the latest showing an improvement up to 13% in yield estimation accuracy.

This work would not have been possible without the contributions of Ricardo Póvoa, Nuno Lourenço, and Ricardo Martins for their support and valuable discussions on circuits and optimization.

Finally, the authors would like to express their gratitude for the financial support that made this work possible. The work developed in this book was supported in part by the Fundação para a Ciência e a Tecnologia (Grant FCT-SFRH/BD/103337/2014) and by the Instituto de Telecomunicações (Research project RAPID UID/EEA/50008/2013 and UID/EEA/50008/2019).

Lisbon, Portugal   António M. L. Canelas
                                                                Jorge M. C. Guilherme
                                                                   Nuno C. G. Horta

Contents

Abbreviations

AC	Alternate current
ACO	Ant colony optimization
ADC	Analog-to-digital converter
ADE	Analog design environment
ADS	Advanced design system
AIDA	Analog integrated circuit design automation
AMG	Analog module generator
AWE	Asymptotic waveform evaluation
BMF	Bayesian model fusion
BV	Basic variables
CAGR	Compound annual growth rate
CMOS	Complementary metal-oxide-semiconductor
CW	Cloud width
DAC	Digital-to-analog converter
DC	Direct current
DE	Differential evolution
DOE	Design-of-experiments
EA	Evolutionary algorithms
EDA	Electronic design automation
FCM	Fuzzy C-means
FoM	Figure-of-merit
FUZYE	Fuzzy c-means based yield estimation
GA	Genetic algorithm
GBW	Gain-bandwidth product
GDC	Gain DC
GP	Geometric programming
GSA	Gravitational search algorithm
GUI	Graphical user interface
HAC	Hierarchical agglomerative clustering

HAD	Hierarchical analog design
HSMC	High-sigma Monte Carlo
IBS	Importance boundary sampling
IBY	Individual-based yield
IC	Integrated circuit
ICs	Integrated circuits
IRDS	International roadmap for devices and systems
IS	Importance sampling
ISE	Infeasible solution elimination
KMD	K-Medoids
KMS	K-Means
LAA	Linear assignment algorithm
LDS	Low discrepancy sequence
LHS	Latin hypercube sampling
LNA	Low noise amplifier
LP	Linear programming
MADS	Mesh adaptive direct search
MC	Monte Carlo
MCS	Monte Carlo pseudo-random sampling
MOEA/D	Multi-objective evolutionary algorithm based on decomposition
MOPSO	Multi-objective particle swarm optimization
MOSA	Multi-objective simulated annealing
MPI	Message passing interface
NBV	Non-basic variable
NMOS	n-Channel MOSFET
NSGA-II	Nondominated sorting genetic algorithm-II
OAD	Orthogonal array design
OCBA	Optimal computing budget allocation
OO	Ordinal optimization
OpAmp	Operational amplifier
OPDK	Organic process design kit
ORDE	Optimization-based random-scale differential evolution
OTA	Operational transconductance amplifier
OTFT	Organic thin-film transistors
PAD	Procedural analog design
PC	Partition coefficient
PCA	Principal component analysis
PDF	Probability density function
PDK	Process design kit
PDKs	Process design kits
PE	Partition entropy
PLL	Phase locked loop
PMOS	p-Channel MOSFET
POF	Pareto optimal front

PRSA	Parallel recombinative simulated annealing
PSA	Pattern search algorithm
PSO	Particle swarm optimization
PVT	Process voltage and temperature
QMC	Quasi-Monte Carlo
RF	Radio frequency
RSM	Response surface methodology
s.t.	Subject to
SA	Simulated annealing
SBX	Single binary crossover
SoC	System-on-chip
SPS	Stochastic pattern search
SQP	Sequential quadratic programming
SR	Sparse regression
SSS	Scaled sigma sampling
SVM	Support vector machine
UD	Uniform design
VLSI	Very large-scale integration
WCD	Worst-case distance
WCP	Worst-case performance
WCPF	Worst-case pareto front
XML	Extensible markup language

List of Figures

List of Tables

Chapter 1
Introduction

1.1 Variability Effects in Analog IC

The increased complexity in today's integrated circuits (ICs), where very large scale integration (VLSI) technologies progressed towards mixed-signal ICs, having digital and analog circuits coexisting on the same die as a complete system-on-a-chip (SoC) [1], allied to the adoption of smaller nanometer-scale integration technologies, creates new challenges to robust ICs design. Although the analog section occupies a small area in the entire chip, the analog design effort is considerably higher than the design of the digital blocks [2]. In traditional analog design, the designer achieves a valid circuit design configuration based on its knowledge and assisted by proper tools, like electrical simulators to validate the required specifications. This is usually a very time-consuming and error-prone iterative process, where a large number of specification constraints must be satisfied, like minimum DC gain, phase margin, and area. In today's competitive electronic market it is not enough to perform a basic circuit sizing process where only a feasible solution is found, i.e., a circuit sizing solution that fulfills all the required specification. Nowadays several of the specifications must be optimized, like power consumption and/or DC gain, which increases the complexity of manual design. Performing space exploration to achieve optimal circuit sized solutions has become a very hard task to IC designers. The increased number of variables and the highly nonlinear relation between circuit design variables and performance specifications brought by new smaller technology nodes made the use of an automatic circuit-level sizing and optimization tool, a requirement for satisfying the time-to-market pressure to release new and high-performance ICs.

Having automatic tools that can quickly provide a sized circuit solution or a set of solutions, when more than one objective is being optimized, is not enough. The economic pressure to produce affordable electronic devices revealed the need to fabricate more reliable circuit solutions. The inherently stochastic nature of semiconductor manufacturing processes led to the appearance of yield losses. The yield

© Springer Nature Switzerland AG 2020   1
A. M. L. Canelas et al., *Yield-Aware Analog IC Design and Optimization in Nanometer-scale Technologies*, https://doi.org/10.1007/978-3-030-41536-5_1

losses in silicon wafers that survive to production can be classified as *catastrophic* or *parametric* [3]. The catastrophic yield losses refer to functional failures, where the circuits do not work at all. These failures may be caused by short or open circuits. The parametric yield losses are caused by random and undesirable variations in circuit parameters due to nonideal fabrication processes, which lead to working and functional circuits that fail to comply with the required specifications. To improve productivity and avoid expensive redesign cycles it is important to predict the circuit parametric yield value during early stages of the circuit sizing processes. The parametric yield value refers to the percentage of circuits that are expected to comply with the required circuit specifications after fabrication. Technology scaling, the appearance of new materials and devices, combined with more demanding operating conditions, e.g., extreme temperatures and high radiation levels, poses new challenges to parametric yield estimation.

For several years, analog integrated circuit (IC) designers considered only global variations, or inter-die variations, and simulated the effects of these variations on the circuit performance by using corner analysis, which works well in digital project, but it is not enough to ensure that analog circuit performances are met after chip fabrication [4]. Analog ICs are also particularly sensitive to local variations, like devices mismatch effects, especially in the nanometer-scale integration technologies. Traditionally, the effects of local variations were prevented and corrected by the experience and know-how of the designer, which may lead to very conservative designs. In Table 1.1 it is possible to observe how the scaling effects impact the variability of the transistor threshold voltage standard deviation $\sigma(V_{Th})$ normalized by the threshold voltage for several nanometer technology nodes.

As a result of the increased impact of the variability effects, several techniques have been proposed to estimate the parametric yield. These techniques can be classified into two main categories: Monte Carlo (MC) based and performance model based.

MC analysis is considered the gold standard for parametric yield prediction, since it is the most reliable and accurate method to estimate the circuit parametric yield [6]. MC analysis performed in electrical simulators is based on statistical device models developed and tested by the technologies' foundries, which typically includes global and local variations. The main drawback of the MC approach is related to the high number of simulations needed in order to provide an accurate parametric yield estimation. The considerable amount of time needed to perform those simulations represents a huge obstacle to the adoption and integration of this type of approach in an electronic design automation (EDA) tool, since it would represent a severe bottleneck in the overall circuit synthesis process. In spite of all the drawbacks, the MC simulation-based yield estimation high accuracy keeps driving research in order to reduce its computational burden, which allows its adoption inside a yield-aware circuit sizing and optimization loop.

Table 1.1 Intra-die variability increase for the transistor threshold voltage parameter with CMOS technology node [5]

Technology node	250 nm	180 nm	130 nm	90 nm	65 nm	45 nm
$\sigma(V_{Th}\,(mV))/V_{Th}\,(mV)$	21/450 = 4.7%	23/400 = 5.8%	27/330 = 8.2%	28/300 = 9.3%	30/280 = 10.7%	32/200 = 16%

1.2   Work Motivation

The International Roadmap for Devices and Systems 2017 (IRDS:2017) [7] identifies scaling as the first reason for the increasing reliability issues in new ICs
technology integration nodes. Another identified cause for poor circuit reliability is
premature aging, due to the long operation working cycles that electronics devices
are subject in today's applications. To address this problem IRDS:2017 defends the
need to investigate and develop new models, both statistical models of lifetime
distributions and physical models of how lifetime depends on stress, geometries,
and materials. The IRDS:2017 also points out the need for developing new reliability
software tools with capabilities to predict and quantify the effects of variability
during the design process. In addition to the IRDS:2017 concerns, Mladen Nizic,
product marketing director at Cadence©, noted that "Advanced process nodes
typically introduce more parametric variation, . . ., and other manufacturing effects
affecting device performance, making it much harder for designers to predict circuit
performance in silicon. To cope with these challenges, designers need automated
flow to understand impact of manufacturing effects early, . . ." [8].

Analog ICs are expected to have the strongest relative growth of the IC market for
the next five years. Power management, signal conversion, and automotive-specific
analog devices will drive a compound annual growth rate (CAGR) of 6.6%, from
$54.5 billion in 2017 to $74.8 billion in 2022, according to IC Insights [9]. In
Fig. 1.1, the forecast CAGR of different product categories is compared with the
expected CAGR of the total IC market growth of 5.1%. In order to keep driving this
amazing growth, a considerable amount of time and work must be dedicated to
improving the analog design flow and analog EDA tools.

The expected growth of the analog IC market and the challenging demand for
new analog EDA tools, including early variability effects prediction in the design

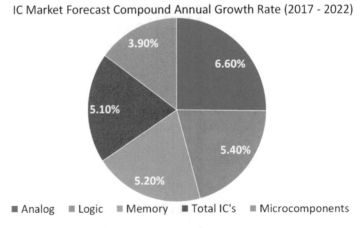

Fig. 1.1 Major IC categories market forecast compound annual growth rate for 2017–2022
(according to [9])

flow, are some of the major motivations for this work. Most of today's circuit sizing EDA optimization tools only performs some type of variability analysis at the end of the optimization process to validate solutions for yield requirements. This type of circuit sizing and optimization flow reduces the time impact of performing variability analysis at each optimization iteration but increases the probability of adding redesign iterations into the sizing processes. The development of a methodology using MC analysis for accurate yield estimation with reduced time impact in the sizing and optimization flow, although challenging, is a requirement to improve ICs reliability and cost efficiency in the manufacturing processes, especially at the new nanometer-scales technology integration nodes.

1.3   Work Purpose

The main goal of the work presented in this book is to develop an accurate yield estimation methodology in order to improve the robustness of circuit solutions sized in a state-of-the-art EDA tool. By improving solutions robustness, the analog IC production processes become more efficient, since expensive redesign iterations are avoided. The EDA tool considered for this work is a circuit sizing design automation solution known as AIDA-C [10–12]. The AIDA-C circuit sizing tool is a simulation-based, multi-constraint, and multi-objective optimization tool, whose optimization kernel is based on evolutionary techniques. Since MC-based approaches are considered the most accurate methods for yield estimation, this work develops its new methodology based on MC analysis. To further improve the yield estimation accuracy, MC analysis will be based on electrical simulations using the trustworthy industry process design kits (PDKs) statistical models.

Combining MC analysis for yield estimation with a population-based evolutionary optimization algorithm, where each population individual is evaluated through electrical simulation, may represent a huge bottleneck in the overall optimization process. Reducing the time impact caused by MC simulations is possible by performing less simulations or iterations, as many electrical simulators call them, to each individual in the population. Another approach is selecting some relevant individuals from the population and performing MC analysis only to those individuals. Since the approach of performing less simulations would reduce the yield estimation accuracy, this type of approach was discarded.

According to the requirements presented so far, the new yield estimation methodology must satisfy the following conditions:

1. Adopt MC analysis to estimate the yield.
2. Perform MC simulations in a commercial electrical simulator with standard PDKs statistical models, for more accurate results.
3. Completely integrate a yield estimation methodology in the optimization loop of a population-based optimization analog IC sizing tool.

4. Reduce the time impact of MC simulations by performing simulations to a small number of individuals of the population.
5. Develop a technique to identify and select the individuals to perform the MC simulation.
6. Use the simulated individuals yield values to estimate the yield for the rest of the population.

The innovative yield estimation methodology will be validated by performing several circuit sizing and optimization problems in nanometer-scale IC technologies. The obtained results must be compared to the exhaustive approach, where all individuals of the population are subject to MC analysis, to assess the effective reduction of MC simulations.

This work only addresses operating conditions and device process-induced variations with impact in circuit performances, already available and modelled in foundry PDKs. Another important reliability issue in analog ICs, especially at nanometer-scale technologies, is the premature aging of circuits, which cause circuit performance degradation overtime resulting in shorter operational lifetime. The aging topic is not addressed in this work because aging-related variation models are still in development, and according to the IRDS:2017, "It also seems likely that there will be less-than-historic amounts of time and money to develop these new reliability capabilities."

1.4 Book Structure

The presented book is organized as follows:

- Chapter 2 presents the analog IC sizing problem and provides background on automatic analog IC sizing by introducing two different approaches, namely knowledge-based and optimization-based approaches. Circuit design and performance parameters are also presented to explain the concept of feasibility regions, and the relation between circuit design parameters and performance parameters. Additionally, the variability effects are discussed, and the parametric yield is presented.
- Chapter 3 provides background information of different techniques to estimate parametric yield and discusses several state-of-the-art approaches for automatic analog IC sizing and optimization using yield to improve solutions robustness. Several yield estimation techniques adopted by commercial electronic design automation tools are also presented.
- Chapter 4 presents the new methodology for MC-based yield estimation using clustering algorithms to reduce the number of expensive MC simulations. Since clustering algorithms play an important role in the new yield estimation methodology, background information about different clustering algorithms is provided. The different techniques to overcome some minor problems detected in early tests are also presented.

- Chapter 5 presents the state-of-the-art automatic analog IC sizing tool, the AIDA-C tool, where the new yield estimation methodology is to be implemented. The chapter starts by introducing the AIDA-C and its graphical user interface. Next, the changes performed by the introduction of the new feature, which allows optimizing the circuit yield solutions, are presented.
- Chapter 6 presents the test case circuits and reveals the result of the tests using the new yield estimation methodology. The results from the different clustering algorithms adopted are also compared in both memory requirements and runtime.
- Chapter 7 presents the conclusion of this work and proposes additional research topics for future work.

References

1. G. Gielen, Design tool solutions for mixed-signal/RF circuit design in CMOS nanometer technologies, in *Proc. ASP-DAC'07 Des. Automat. Conf.*, 2007
2. M. Barros, J. Guilherme, N. Horta, *Analog Circuits and Systems Optimization Based on Evolutionary Computation Techniques* (Springer-Verlag, Berlin, 2010)
3. P. Gupta, E. Papadopoulou, Yield analysis and optimization, in *Handbook of Algorithms for Physical Design Automation*, (CRC Press, Boca Raton, FL, 2008), pp. 771–790
4. C. McAndrew, I.-S. Lim, B. Braswell, D. Garrity, Corner models: inaccurate at best, and it only gets worst..., in *Proc. IEEE Custom Integr. Circuits Conf.*, 2013
5. C. Chiang, J. Kawa, *Design for Manufacturability and Yield for Nano-Scale CMOS* (Springer, Dordrecht, 2007)
6. A. Singhee, R.A. Rutenbar, Why quasi-Monte Carlo is better than Monte Carlo or Latin hypercube sampling for statistical circuit analysis. IEEE Trans. Comput. Aided Des. Integr. Circuits Syst. **29**(11), 1763–1776 (2010)
7. IEEE IRDS, The International Roadmap for Devices and Systems: 2017, IEEE, 2017
8. G. Moretti, Senior Editor Chip Design Magazine, Complexity of Mixed-Signal Designs, 28 Aug 2014. [Online]. Available: http://chipdesignmag.com/sld/blog/2014/08/28/complexity-of-mixed-signal-designs/. Accessed 16 Sept 2018
9. IC Insights, Analog IC Market Forecast With Strongest Annual Growth Through 2022, 12 Jan 2018. [Online]. Available: http://www.icinsights.com/data/articles/documents/1036.pdf. Accessed 16 Sept 2018
10. N. Lourenço, R. Martins, N. Horta, *Automatic Analog IC Sizing and Optimization Constrained with PVT Corners and Layout Effects* (Springer International Publishing, Cham, 2017)
11. R. Lourenço, N. Lourenço, N. Horta, *AIDA-CMK: Multi-Algorithm Optimization Kernel Applied to Analog IC Sizing* (Springer International Publishing, Cham, 2015)
12. N. Lourenço, R. Martins, A. Canelas, R. Póvoa, N. Horta, AIDA: layout-aware analog circuit-level sizing with in-loop layout generation. Integration VLSI J. **55**, 316–329 (2016)

Chapter 2
Analog IC Sizing Background

2.1 Analog IC Sizing

Circuit sizing is the process of finding the appropriate physical parameters of every device in a circuit topology, such that the circuit is able to comply with the required performance specifications. In a circuit such as the two-stage Miller amplifier example in Fig. 2.1, the parameters that must be set to achieve the desired performance specifications detailed in Table 2.1 are the gate width, gate length, and number of fingers of the MOS transistors, for the capacitor are the number of fingers and finger length, whereas for the resistor are the width and length of the resistor poly finger.

The simple circuit example in Fig. 2.1 implies finding a total of 12 gate width/length parameters for the MOS transistors and 2 for each passive device in order to achieve the required performance specifications. The number of the transistors parameters is lower than expected because several devices share the same gate width/length due to matching device requirements. Although the matching among several transistors reduce the number of design parameters, it is common to adopt transistors structures with more than one gate finger per transistor, in order to avoid long transistor gates, which results in adding a new design parameter for each transistor, i.e., the number of fingers. Considering the two-stage Miller amplifier a total number of 22 design parameters, detailed in Table 2.2, must be found to achieve a sized solution that comply with the desired performance specifications.

The difficult task of finding the correct value for each design parameter is not only related to the high number of parameters but also because, typically, a trade-off occurs among the different performance specifications; i.e., when one performance measure gets closer to the target desired value, the other(s) may move away from the target value specification, making difficult to satisfy all performance specifications. Additionally to the desired performance specifications are functional specifications that also must be fulfilled, like assuring that transistors are working on the desired operating region, e.g., MOS transistors are on the saturation region, Table 2.3.

© Springer Nature Switzerland AG 2020

A. M. L. Canelas et al., *Yield-Aware Analog IC Design and Optimization in Nanometer-scale Technologies*, https://doi.org/10.1007/978-3-030-41536-5_2

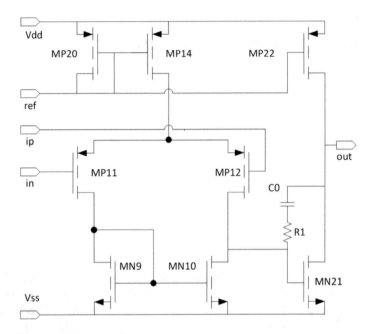

Fig. 2.1 Two-stage Miller amplifier

Table 2.1 Two-stage Miller amplifier desired performances

Performance specification	Target value	
Current consumption	≤200	μA
Low-frequency gain	≥55	dB
Unity gain frequency	≥200	MHz
Phase margin	≥55	°
Power supply rejection ratio	≥55	dB
Structural offset voltage	≤5	mV
Noise RMS	≤600	μVrms
Noise density	≤50	nV/√Hz

Table 2.2 Two-stage Miller amplifier design parameters

Device	Design parameters	Device	Design parameters
Capacitor C0	Length; Number of fingers.	PMOS transistor MP14	Gate length; Gate width; Number of fingers.
PMOS transistor MP20	Gate length; Gate width; Number of fingers.	PMOS transistor MP22	Gate length; Gate width; Number of fingers.
PMOS transistor MP11 (Matched with MP12)	Gate length; Gate width; Number of fingers.	NMOS transistor MN9 (Matched with MN10)	Gate length; Gate width; Number of fingers.
NMOS transistor MN21	Gate length; Gate width; Number of fingers.	Resistor R1	Length of finger; Width of finger.

Table 2.3 Two-stage Miller amplifier functional specifications

Functional specification	Target value (mV)	Functional specification	Target value (mV)
Overdrive voltage MP11 $(V_{TH} - V_{GS})$	≥ 100	Saturation margin MP11 $(VD_{Sat} - V_{DS})$	≥ 50
Overdrive voltage MP12 $(V_{TH} - V_{GS})$	≥ 100	Saturation margin MP12 $(VD_{Sat} - V_{DS})$	≥ 50
Overdrive voltage MP14 $(V_{TH} - V_{GS})$	≥ 100	Saturation margin MP14 $(VD_{Sat} - V_{DS})$	≥ 50
Overdrive voltage MP22 $(V_{TH} - V_{GS})$	≥ 100	Saturation margin MP22 $(VD_{Sat} - V_{DS})$	≥ 50
Overdrive voltage MN21 $(V_{GS} - V_{TH})$	≥ 100	Saturation margin MN21 $(V_{DS} - VD_{Sat})$	≥ 50

V_{TH} transistor threshold voltage, V_{GS} voltage between transistor gate and source terminals, $VD_{Sat} = V_{GS} - V_{TH}$ for NMOS transistors, $VD_{Sat} = V_{TH} - V_{GS}$ for PMOS transistors, V_{DS} voltage between transistor drain and source terminals

Achieving a circuit sizing solution where a large number of specifications must be fulfilled is a very difficult task, which is why designers define a class of *weak* specifications where noncritical specification target values are relaxed; for instance, in the two-stage Miller amplifier circuit the low-frequency gain is relaxed to a smaller value of 50 dB, to easily fulfill this specification. Whereas critical specifications, such as specifications assuring circuit stability, like phase margin, are labelled as *strong* specifications and no relaxation is allowed [1]. The adoption of new nanometer technology nodes and device structures, such as finFET devices [2, 3], allow a higher transistor density but led to the appearance of new strong specifications making very difficult for designers to achieve a circuit sized solution, which boost the development of automatic circuit sizing tools.

2.2   Automatic Analog IC Sizing

The development of mature analog EDA tools [4], which had always followed the steps of the highly developed EDA tools for its digital counterpart, is the key to deal with the increased analog IC design complexity. Analog circuit design is highly based on the well-kept knowledge and experience of designers, which is why analog design is often referred as an art or even a "black art" [5]. The absence of a well-defined detail design flow, generally accepted and adopted by all designers and companies, is one of the main difficulties to achieve a complete analog EDA tool [6]. Nevertheless, it was possible to identify common tasks performed by every designer, which made possible to develop an analog design flow. In Fig. 2.2a the hierarchical analog design flow, introduced by Gielen and Rutenbar [7], is presented. This hierarchical design flow made possible to systematize the design process at different levels of abstractions, from system level design to physical layout, allowing an early identification of errors, thus avoiding redesign iterations and increasing the chance of first-time-right design.

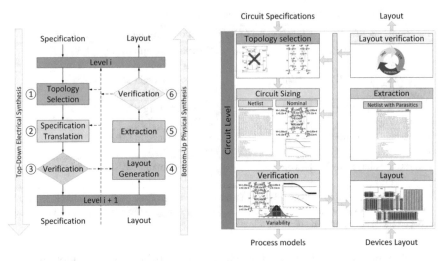

Fig. 2.2 (**a**) Hierarchical analog design (HAD) methodology. (**b**) Typical HAD implementation at circuit level

The Hierarchical Analog Design (HAD) methodology, in Fig. 2.2a, which is adopted by most EDA tools, decomposes into smaller and manageable blocks, a more complex circuit, e.g., an RF receiver into filters, amplifiers, and so forth. The new smaller sub-blocks can be autonomously designed, at lower levels in the hierarchy, using legacy specifications from higher levels.

The adoption of HAD makes possible to repeat the decomposition process until a level of physical implementation is reached. After the top-down decomposition process (steps 1–3) a bottom-up layout process starts (steps 4–6). If at some level a desired specification is impossible to achieve, a redesign iteration is required. Typical levels defined in HAD are *system, architecture, circuit, device,* and *process.* At *circuit* level is where the circuit sizing task occurs. The circuit is defined by means of a connected devices netlist. The adopted circuit devices invoke device models defined in library files, selected at a lower hierarchy *device* level. In turn, the device model's library is dependent on the manufacturing process selected at *process* level.

In Fig. 2.2b, a typical HAD at circuit level is detailed. The main tasks of the HAD *circuit* level are described as follows:

– *Circuit topology selection*, as the name suggests, is where the most adequate circuit topology is chosen according to circuit performance requirements and specifications. This selection process is, typically, knowledge based and it is achieved by comparing the required circuit specifications with the characteristic of each topology in terms of achievable specifications. This task is usually performed by designers, where their knowledge and experience about the differ- ent topology's characteristics is the key for a successful choice. Nowadays, automatic topology selection tools already exist. Most of those topology selection tools are limited to classes of circuits, like amplifiers [8]. Those tools characterize circuit topologies in terms of potential achieved specifications, either by solving circuit equations or by performing a large number of electrical simulations. Next, using this information performs the selection using rule-based filtering [9] or searching in a tree structure if desired specifications are satisfied [10].

- *Circuit sizing* task is, typically, an iterative process where circuit parameter values, such as CMOS transistors gate widths/lengths, resistors or capacitors values and bias voltages/currents, are defined. After selecting the topology and having defined the devices' models according to the selected technology node, the circuit netlist that defines how the circuit devices are interconnected is created. Also, in this netlist file the performance measures to be computed by the electrical simulator are coded, if simulation-based tools are adopted in the sizing processes. Circuit sizing is done in two steps [11]. The first step adopts typical process parameters values, where no variability effects are considered. The second step is performed in the verification stage, where variability conditions are introduced in the sizing processes in order to improve the solutions robustness. Analog circuit parameter values search is severely conditioned by a large number of constraints, imposed by the required specifications and/or by some functional conditions, like having all transistors operating in the saturation region. Dealing accurately and effectively with a large number of constraints that must be fulfilled and performing an efficient parameter variable space exploration, to improve several key circuit specifications, is one of the main reasons for developing automated circuit sizing methodologies.

- *Circuit Verification* validates circuit performances under variability conditions. This task considers variability device models and worst-case conditions in simulations to check if circuit performance requirement still hold under these conditions [12, 13]. In case some performance requirement fails, then a second sizing step is performed under variability conditions to obtain a solution with enough safety margins to hold under variability conditions.

- *Layout* task is where devices physical representation is laid or placed in the circuit floor plan and devices interconnection is performed according to the selected schematics topology, Fig. 2.3 [14, 15]. The layout is used to create the different masks that enabled the real physical chip fabrication.

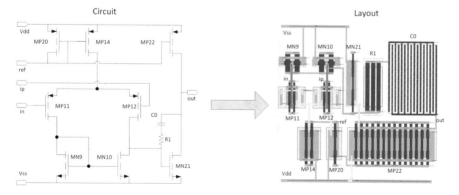

Fig. 2.3 IC layout design

- *Extraction* produces a netlist which includes parasitic effects introduced by the layout [16]. Those effects can severely degrade circuit performances or even cause circuit malfunction.
- *Layout verification* performs a set of different validation steps on the layout design, such as checking if the desired performance requirements hold under parasitic effects, by using the extracted netlist with parasitics, also design rules check and layout versus schematic validation steps are performed [17].

The complexity of the analog IC sizing task, brought by the new nanometer technology nodes, pointed out the importance of developing new and better sizing strategies. The evolution of automatic analog IC sizing tools was based on two different approaches, namely knowledge-based and optimization-based approaches [18].

2.2.1 *Knowledge Based*

In knowledge-based approach, a manual design plan, based on the expert knowledge of a designer's team, is developed. The design plan captures designers' expertise in a set of equations that relate circuit performance to device characteristics, which are explicitly solved to obtain a circuit sizing solution where the performance requirements are fulfilled, Fig. 2.4.

In the 1980s decade and early 1990s, EDA tools adopted the knowledge-based approach. In IDAC [19], after choosing the circuit topology, the design plan is executed for a set of desired specifications, which correspond to solve all design equations for a particular set of given specifications. This tool offers a library of

Fig. 2.4 Knowledge-based approach

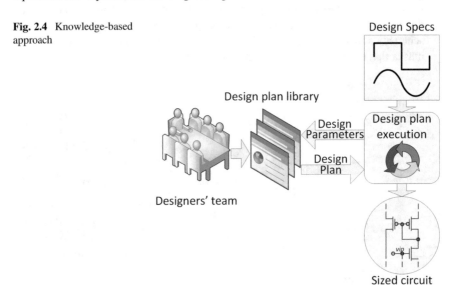

preconfigured design plans based on different types of circuits, such as voltage references, OpAmps, comparators, oscillators, DACs, and ADCs.

The OASYS [20] approach adopts the same overall strategy. OASYS moves into a more detailed design plan by dividing the circuit into sub-blocks and executing a design plan for each circuit sub-block, e.g., current mirror, current source, and simple differential pair. At higher levels of abstraction, the TAGUS [21–23] approach performs the synthesis process by determining the circuit appropriate functional building blocks and by using knowledge-based rules to instantiate such building blocks by electrical sub-circuits. The Procedural Analog Design (PAD) [24, 25] tool also identifies basic circuit blocks, where the designer is able to modify interactively, block by block, a subset of the circuit design parameters and observes the effect on the overall circuit parameters.

In order to improve process automation, works such as [26–28] implement expert systems using artificial intelligence techniques to capture circuit knowledge. The BLADES [26] approach adopts a divider and conquer strategy using different levels of abstraction where design primitives are defined, which allows to easily extract the design plans.

Typically, knowledge-based approaches are used to achieve first-cut-solutions, where the tool focus is on obtaining a simple working circuit solution. Most recent works try to overcome the first-cut-solution idea by adopting hybrid approaches, where a knowledge-based technique is used to fast obtain an approximate solution, followed by an optimization process which fine tunes the circuit design parameters. An example of this hybrid approach is used in [29], where Genetic Algorithm (GA) is adopted as the optimization kernel.

The main advantage of knowledge-based approaches is the short execution time to obtain a feasible sized circuit. The difficulty in obtaining and maintaining design plans, since it is a very time-consuming task, is one of the major limitations of this approach. Other problems for the adoption of knowledge-based approaches are related to accuracy, because design rules in the form of circuit analysis equations are often subject to considerable simplifications, which is why this type of approach is mainly used with success in "older" and well-established technologies, where those simplifications are still able to provide a sufficient accurate feasible design plan and solution.

2.2.2 Optimization Based

The demanding for high-performance analog ICs boosts the search and development of new automatic circuit sizing techniques. The optimization-based approach, Fig. 2.5, has emerged from this search and is now the most common approach to automatic circuit sizing tools. This approach was also adopted by the main commercial EDA providers, such as Cadence in the Analog Design Environment XL and GXL [30], MunEDA WiCkeD [31], and Magma Titan ADX [32] acquired by Synopsys in 2012 [33].

Fig. 2.5 Optimization-
based approach

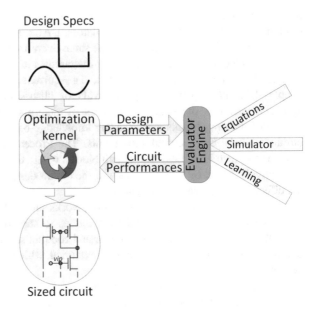

Design Specs

Optimization kernel

Design Parameters

Circuit Performances

Evaluator Engine

Equations

Simulator

Learning

Sized circuit

The optimization-based approach converts the circuit sizing process into a single-
or multi-objective constrained optimization problem as expressed in (2.1).

$$
\begin{aligned}
&\min \ f_i(\mathbf{x_d}) \quad i = 1, 2, \ldots, \quad \#\text{Objectives} \\
&\text{subject to} \\
&\quad g_j(\mathbf{x_d}) \leq 0, \quad j = 1, 2, \ldots, \quad \#\text{Inequalities} \\
&\quad h_k(\mathbf{x_d}) = 0, \quad k = 1, 2, \ldots, \quad \#\text{Equalities} \\
&\quad \mathbf{x_d} = [x_{d,1} \ldots x_{d,n}]^T \in \mathbb{R}^n, \quad X_{\mathrm{L}}^{(m)} \leq x_{d,m} \leq X_{\mathrm{U}}^{(m)}, \quad m = 1, 2, \ldots, n
\end{aligned}
\tag{2.1}
$$

A solution of the optimization problem (2.1) represents a possible combination of
design variables parameters $\mathbf{x_d}$, e.g., transistors' gate width/length or circuit bias
currents, such that the sized circuit meets all the desired performance specifications.
Each component $x_{d,\,m}$ of vector $\mathbf{x_d}$ corresponds to a circuit design variable usually
bounded by a lower value $X_{\mathrm{L}}^{(m)}$ and an upper value $X_{\mathrm{U}}^{(m)}$. Often design variables'
ranges are derived from technological restrictions, such as minimum transistor gate
width and length, while others result from electrical constraints imposed by surrounding circuits that will interact with the circuit under optimization. $f_i(\mathbf{x_d})$ represents the cost or objective functions to be optimized, which usually corresponds to
some important circuit performance requirements that the designer wants to
improve, e.g., DC gain, bandwidth, and power consumption. Although in (2.1) the
optimization problem is defined as a minimization problem, it is also possible for the
maximization of any objective by simply transforming maximization into minimi-
zation by inverting the sign of the function(s) $f_i(\mathbf{x_d})$. The functions $g_j(\mathbf{x_d})$ and $h_k(\mathbf{x_d})$
refer, respectively, to inequalities and equality constraints. The constraint functions

Fig. 2.6 Pareto front illustration

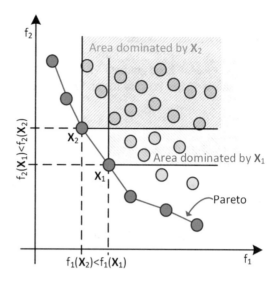

impose some functional/performance conditions to which a solution must comply, such as minimum phase margin, DC gain or minimal saturation margin in the transistors. Several functional constraints can automatically be derived from sizing rules as described in [34, 35].

In circuit sizing and optimization problems where the designer only wants to improve one performance figure, i.e., single objective, the optimization algorithm used to solve the problem will reach, at best, only one optimal sizing solution. In the case of having more than one cost function, which corresponds to a multi-objective optimization problem, the final trade-off attained in the solutions defines a Pareto optimal front (POF), Fig. 2.6.

In multi-objective optimization problems where the goal is the minimization of the problem objectives, the nondominated solutions defining the POF are selected based on a criterion of dominance defined as follows: point $\mathbf{X}_1$ is not dominated by another point $\mathbf{X}_2$, if $\exists_i : f_i(\mathbf{X}_1) < f_i(\mathbf{X}_2)$, with $i = 1, 2, \ldots, \#$ Objectives.

2.2.2.1 Optimization Techniques

Before discussing several circuit sizing and optimization-based works and to provide some background about the adopted optimization techniques in those works, a brief description of those techniques is presented. For further reading, in [36] a brief and interesting historical perspective in optimization is presented.

Steepest Descent Optimization

This method is also known as gradient descent optimization, and was proposed by Cauchy in 1847. It is an iterative method for finding the minimum of a function that starts at an initial solution, and evolves the solution according a search direction defined by a vector based on the negative value of the gradient of the objective function, with steps proportional to the referred gradient. Consider function $f(\mathbf{x}) : \mathbb{R}^n \rightarrow \mathbb{R}$, whose gradient $\nabla f(\mathbf{x})$ is possible to compute and an initial point $\mathbf{x}_0$, the iterative update process for finding the minimum point is defined by:

$$\mathbf{x}_{k+1} = \mathbf{x}_k - \gamma \nabla f(\mathbf{x}_k) \tag{2.2}$$

where γ defines the step length and must be small enough to ensure convergence of the method. Selecting a very small value for γ results in a very long execution time, which is why, typically, the step value is adjusted at each iteration of the algorithm. Several works have addressed the problem of finding the optimal step size [37, 38].

Algorithm 2.1 Steepest Descent

Given
    $\mathbf{x}_0$ initial solution,
    $g_0 = \nabla f(\mathbf{x}_0)$ gradient of f in $\mathbf{x}_0$,
    tol convergence tolerance parameter,
    maxiter maximum number of iterations.
$k \leftarrow 0$
While $k < maxiter$ do
    find optimal γ_k
    $\mathbf{x_{k+1}} \leftarrow \mathbf{x_k} - \gamma_k \ \nabla f(\mathbf{x_k})$
    compute $g_{k+1} = \nabla f(\mathbf{x_{k+1}})$
    if $\| g_{k+1} \|_2 \leq tol$ then
            converged: exit and return solution
    $k \leftarrow k+1$
End While
Did not converged.

Due to its greedy nature in the look for the optimum solution and depending on the initial point $\mathbf{x}_0$ solution, the steepest descent method can get trapped at different local minimum solutions. This method only assures getting the optimum solutions for convex functions, $f_1(x)$ (2.3), Fig. 2.7, and for strictly convex functions, $f_2(x)$ (2.4), Fig. 2.8, at most one solution is found.

$$f_1(x) : X \rightarrow \mathbb{R} \text{ is convex if :}$$
$$\forall x_1, x_2 \in X, \ \forall t \in [0, 1] : \tag{2.3}$$
$$f_1(tx_1 + (1 - t)x_2) \leq tf_1(x_1) + (1 - t)f_1(x_2)$$

$$f_2(x) : X \rightarrow \mathbb{R} \text{ is strictly convex if :}$$
$$\forall x_1, x_2 \in X, x_1 \neq x_2, \ \forall t \in [0, 1] : \tag{2.4}$$
$$f_2(tx_1 + (1 - t)x_2) < tf_2(x_1) + (1 - t)f_2(x_2)$$

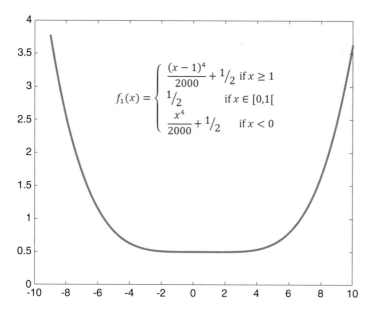

Fig. 2.7   Convex function example

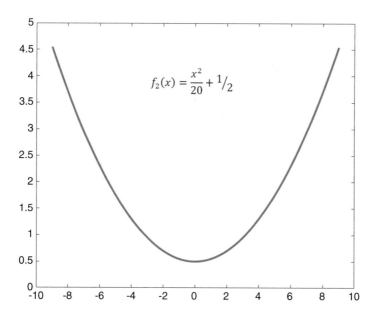

Fig. 2.8   Strictly convex function example

Linear Programming (LP)

The method name derives from the type of optimization problems it solves. To adopt this method, the objective function $f(\cdot)$, inequality constraints $g(\cdot)$, and equality constraints $h(\cdot)$, presented in (2.1), must be linear. In canonical form, LP problems are formulated as:

$$\text{maximize or minimize } \mathbf{c}^T \mathbf{x}$$
$$\text{s.t.}\quad \mathbf{Ax} \le \mathbf{b} \tag{2.5}$$
$$\mathbf{x} \ge \mathbf{0}$$

where s.t. is the acronym to "subject to," $\mathbf{x} \in \mathbb{R}^n$ is a vector of variables. $\mathbf{b}$ and $\mathbf{c}$ are vectors of coefficients and A is a matrix of coefficients.

In canonical form only inequality constraints are accepted; transforming equality constraints is possible by decomposing the equality constraint into two new inequality constraints:

$$\sum_j a_{ij}x_j = b_i \rightarrow \begin{cases} \sum_j a_{ij}x_j \le b_i \\ \sum_j a_{ij}x_j \ge b_i \end{cases} \tag{2.6}$$

where a_{ij} is a coefficient in position ij from matrix A, x_j is variable j and b_i is coefficient i from vector $\mathbf{b}$.

In simple LP problems of two or three dimensions, i.e., $\mathbf{x} \in \mathbb{R}^2$ or $\mathbf{x} \in \mathbb{R}^3$, a graphical method can be applied to find a solution. This simple graphical method plots the convex polytope feasible region, defined by the linear inequality constraints. Then, based on the objective function it is possible to define planes where the function assumes constant values. Moving these planes along its normal vector makes possible to find the optimal solution by computing the function at each position where the plane intersects the feasible space. As an application example of this graphical method, Fig. 2.9, consider the following maximization LP problem:

- Find (x_1, x_2) that maximize the objective function $x_1 + x_2$, subject to $x_1 \ge 0$, $x_2 \ge 0$, $x_1 + 2x_2 \le 4$, $4x_1 + 2x_2 \le 12$, and $-x_1 + x_2 \le 1$.
  The objective function has constant values for lines perpendicular to the vector $(1, 1)$; moving those lines along the vector away from the origin, it is possible to observe that the value of function $x_1 + x_2$ is increasing, so the maximum value for the objective function occurs at the farthest point from the origin where the constant lines intersect the feasibility region, i.e., point (8/3, 2/3).

Dealing with more complex LP problems requires a different strategy. The simplex method was presented in 1947 by George Dantzig. This method defines a standard form to represent the LP problem. In this standard form all the inequality constraints are converted into equality constraints by adding a *surplus* variable,

Fig. 2.9 Linear
programming graphical
method

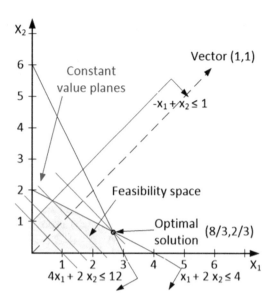

sometimes called *excess* (2.7), or a *slack* variable (2.8). This transformation does not
include the $\mathbf{x} \geq \mathbf{0}$ constraints. Adopting this equation form of representation for the
LP problems reveals much easier to perform algebraic and computational
manipulations.

$$\sum_j a_{ij}x_j \geq b_i \rightarrow \sum_j a_{ij}x_j - e_i = b_i; \quad e_i \geq 0, \; surplus \text{ variable} \qquad (2.7)$$

$$\sum_j a_{ij}x_j \leq b_i \rightarrow \sum_j a_{ij}x_j + s_i = b_i; \quad s_i \geq 0, \; slack \text{ variable} \qquad (2.8)$$

After performing this transformation, the LP problem in standard form representation is given by (2.9):

$$\text{minimize } z = \mathbf{c}^T\mathbf{x}$$
$$\text{s.t.} \quad \mathbf{A}\mathbf{x}^{(\prime)} = \mathbf{b} \qquad (2.9)$$
$$\mathbf{x}^{(\prime)} \geq \mathbf{0}$$

where vector $\mathbf{x}^{(\prime)}$ is an enlarge version of the original $\mathbf{x}$ vector with the new s_i slack
and e_i surplus variables. Also, matrix A now includes the corresponding coefficients
for the slack and surplus variables.

Considering that the system of equations $\mathbf{A}\mathbf{x}^{(\prime)} = \mathbf{b}$ has m equations and n variables, with $n \geq m$, it is possible to obtain a basic solution by setting $n - m$ variables
to zero and solving the system of equation for the rest of the variables. The variables
whose value was set to zero are called nonbasic variables (NBV), and the remaining
m variables are called basic variables (BV). In a sense, the simplex method finds an

optimum set of BV by iteratively exchanging NBV and BV using the pivot operation. The pivot operation swaps the BV with NBV after scaling the row that contains
the BV so that the coefficient of the NBV becomes one. After, multiples of the scaled
row are added to all the other rows so that they have a coefficient of zero in the
column of the NBV [39, 40].

Geometric Programming

Geometric programming (GP) optimization requires that both objective functions
and constraints be formulated in a particular form [41]. In GP standard form, the
optimization problem described in (2.1) must be formulated with objective functions
$f_i(\mathbf{x}_d)$ and inequality constraints $g_j(\mathbf{x}_d)$ as posynomials and equality constraints $h_k(\mathbf{x}_d)$
as monomials. A function $f(\mathbf{x}) : \mathbb{R}^n \to \mathbb{R}$ is a monomial; in the context of GP, it can
be expressed as:

$$f(\mathbf{x}) = cx_1^{\alpha_1} x_2^{\alpha_2} \ldots x_n^{\alpha_n} = c \prod_{k=1}^{n} x_k^{\alpha_k} \tag{2.10}$$

where $c > 0$ and $\alpha_k \in \mathbb{R}$.

A posynomial function is a sum of one or more monomials:

$$f(\mathbf{x}) = \sum_{i=1}^{m} c_i \prod_{k=1}^{n} x_k^{\alpha_{ik}} \tag{2.11}$$

where $c_i > 0$.

Applying to the constraints and objective functions a logarithmic transformation,
$y_k = \log x_k$, is possible to easily convert a GP problem to a nonlinear convex
optimization problem.

$$h(\mathbf{y}) = \log \left(f(e^{y_1}, \ldots, e^{y_n}) \right) = \sum_{i=1}^{m} c_i e^{\left(\sum_{k=1}^{n} y_k \alpha_{ik} \right)} \tag{2.12}$$

Once in convex form, many numerical and traditional algorithms can be used to
solve GP problems. Most state-of-the-art solutions adopt interior-point algorithms
[41]. The main advantage of GP is that it is able to solve large optimization
problems, i.e., thousands of variables and tens of thousands of constraints, very
efficiently in terms of time. Also, a global solution is always found if the problem is
feasible.

Simulated Annealing

The simulated annealing (SA) algorithm [42], as name states, is inspired by the cooling and annealing of liquid materials. By slowly cooling melted materials a state of minimum energy is reached, corresponding to a defect-free crystal structure. SA emulates the cooling process by using a temperature control parameter. Also, the energy of a state represents the objective function and a physical material state corresponds to a solution.

SA is a descent algorithm allowing random ascent moves to avoid getting trapped in local minima solutions. The ascent moves are decided by a probability function based on the energy difference of two states, given by the Boltzmann factor:

$$P = e^{-\frac{\Delta E}{T}} \tag{2.13}$$

where T represents the temperature and $\Delta E = E_2 - E_1$ is the energy difference between two states, or in the case of SA between the objective function values in two iterations.

SA algorithm starts with high temperature values, and for positive energy differences the probability of moving uphill is high. At the initial stage, the algorithm performs a coarse search of the solution space to identify areas where the minimum might be located; this stage is called the exploration phase. As the temperature descends, the probability of ascent moves decrease and a fine search for the minimum starts; this stage is called exploitation phase. The transition between exploration and exploitation phase is critical to the efficiency of SA. Reducing the temperature to fast may result in premature converge to a nonoptimal solution, while for a slow temperature reduction results in long convergence time. Several approaches are adopted for temperature reduction [43], a typical annealing schedule is given by:

$$T(t) = \alpha T(t-1), \quad \text{with } 0.85 \leq \alpha \leq 0.96 \tag{2.14}$$

SA algorithm main steps are outlined in Algorithm 2.2. The algorithm starts at a random initial solution $\mathbf{x_0}$; also the initial parameter T_{max} related to the temperature is set to a high value. To finish the algorithm and exit the *while* loop, a criterion using a minimum temperature, T_{min}, value is defined. The search for the optimal solution is achieved by applying random perturbations, $\Delta \mathbf{x}$, to the current solution, $\mathbf{u}$. The new point $\mathbf{v} \leftarrow \mathbf{u} + \Delta \mathbf{x}$ is accepted as the new solution if the new energy state is lower than the previous, which in the presented algorithm is decided by the negative difference between the objective function computed at the new point $\mathbf{v}$ and the value evaluated at point $\mathbf{u}$. The new solution $\mathbf{v}$ is also accepted if the probability from (2.13), at the current temperature, is higher than some random number obtained from a uniform distribution between zero and one, $U(0, 1)$. The cooling scheduling is represented in the algorithm by a reduction of ΔT of the current temperature T, but this line can be replaced by any other cooling scheduling like in (2.14).

Algorithm 2.2 Simulated Annealing

Given

$\mathbf{x}_0$	initial solution,
T_{max}	initial high temperature,
T_{min}	final low temperature.

$T \leftarrow T_{max}$

$\mathbf{u} \leftarrow \mathbf{x}_0$

While $T > T_{min}$ do

$\quad \mathbf{v} \leftarrow \mathbf{u} + \Delta \mathbf{x}$

$\quad \Delta E = E(\mathbf{v}) - E(\mathbf{u})$

$\quad$ if $\Delta E < 0$ then

$\quad \quad \quad$ accept new solution: $\mathbf{u} \leftarrow \mathbf{v}$

$\quad$ else if $P = e^{-\frac{\Delta E}{T}} > U(0, 1)$

$\quad \quad \quad$ accept new solution: $\mathbf{u} \leftarrow \mathbf{v}$

$\quad T \leftarrow T - \Delta T$

End While

return $\mathbf{u}$.

Stochastic Pattern Search

Stochastic pattern search (SPS) algorithm is based on the Pattern Search algorithm (PSA) formalized by Torczon in [44]. PSA is a greedy optimization method, where the search for the optimal solution points always to the best solution found so far. The basic idea of PSA is similar to SA, but in this case no uphill movements are accepted. PSA starts from an initial solution $\mathbf{x}_0$, which is next perturbed according to a pattern that generates new potential solutions in different directions, which are next evaluated. In case one of the new solutions is better than the initial solution, the algorithm repeats the perturbation process, now considering the new best solution as the central point of the pattern. If all new solutions are worse than the initial point, the step of the pattern is reduced. In Fig. 2.10 this idea is illustrated; the pattern

Fig. 2.10 Search pattern example

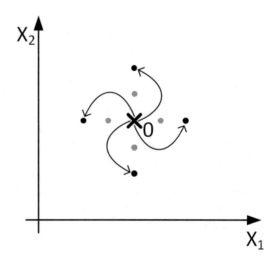

perturbation around solution $\mathbf{x_0}$ results in new potential solutions, represented as black dots. These new solutions are evaluated and if one of them is better than the initial solution, then the pattern moves to that new solution and applies a new perturbation. In case none of the new solution is better than $\mathbf{x_0}$, the grid step of the pattern is reduced, and a new pattern perturbation is generated around the initial solution, creating a set of new potential solutions, represented as gray dots. This process is repeated until convergence is reached.

Like most greedy algorithms, PSA can get trapped in local optimum solutions. To overcome this problem, SPS [45] applies random perturbations at each variable and, also, the order of the perturbed variables is random, which results in different search pattern for each iteration of the algorithm. Like in PSA, if no better solution is found at some iteration, the perturbation step is decreased.

Particle Swarm Optimization

Particle swarm optimization (PSO) technique was originally developed by Kennedy and Eberhart in [46]. In PSO each potential solution, $\mathbf{x}$, is represented by a particle, and finding the optimal solution is achieved by mutually updating each particle position and a velocity vector that define how a particle moves along the search space. New particle positions are defined by:

$$\mathbf{x}_i(t+1) = \mathbf{x}_i(t) + \mathbf{v}_i(t+1) \tag{2.15}$$

where $\mathbf{x}_i(t)$ represents solution i position at iteration t and $\mathbf{v}_i(t+1)$ is a vector that enforces the particle trajectory and is defined by:

$$\mathbf{v}_i(t+1) = \mathbf{v}_i(t) + c_1(\mathbf{p}_i - \mathbf{x}_i(t))\mathbf{R}_1 + c_2(\mathbf{g} - \mathbf{x}_i(t))\mathbf{R}_2 \tag{2.16}$$

The velocity vector $\mathbf{v}_i(t+1)$ is defined by different terms; the $\mathbf{v}_i(t)$ term represents *inertia* and ensures that the particle follows the previous direction to avoid sudden changes in the search trajectory; the second term $c_1(\mathbf{p}_i - \mathbf{x}_i(t))\mathbf{R}_1$ represents a memory term pointing in the direction of the best positions $\mathbf{p}_i$ that particle i has visited so far and it is often called the *cognitive component*; the last term, called *social component*, represents the best global position $\mathbf{g}$ so far, considering all particles in the swarm. The constant coefficients c_1 and c_2 are weight coefficients that define the contribution of moving towards each best position, $\mathbf{p}_i$ and $\mathbf{g}$, in the velocity vector. Typically, these coefficients assume values between zero and two [47]. $\mathbf{R}_1$ and $\mathbf{R}_2$ are diagonal matrices of random numbers from a uniform distribution $U(0, 1)$. The semi-random movement added by both matrices helps solution space exploration and avoids premature convergence. The PSO runs until some stopping criterion is met, e.g., maximum number of iterations is reached or no improvements in the objective function during several iterations.

Algorithm 2.3 Particle Swarm Optimization Algorithm

Initialization
   Create *N* particles
   BestFit← Sets best global fitness value to a bad value in terms of the optimization goal.
   For *i*=1 to N
      $\mathbf{x}_i(0)$ ← *Rand()*   Sets random initial position
      p_i← $\mathbf{x}_i(0)$ Sets particle best position to its initial position

      if *fitness*($\mathbf{x}_i(0)$) Is Better Than *BestFit* then
         BestFit←*fitness*($\mathbf{x}_i(0)$)
         g← $\mathbf{x}_i(0)$

   End For

t ←0
While *Not StoppingCriterionMet* do
   For *i*=1 to N
      Update velocity vector
      $v_i(t + 1) = v_i(t) + c_1(p_i - \mathbf{x}_i(t))R_1 + c_2(g - \mathbf{x}_i(t))R_2$
      Update particle position
      $\mathbf{x}_i(t + 1) = \mathbf{x}_i(t) + v_i(t + 1)$

      Update best positions
         If *fitness*($\mathbf{x}_i(t+1)$) Is Better Than *fitness*(p_i) then
      p_i← $\mathbf{x}_i(t+1)$
         If *fitness*($\mathbf{x}_i(t+1)$) Is Better Than *fitness*(g) then
                  g← $\mathbf{x}_i(t+1)$
      t←$t+1$
   End For
End While

   return g.

Ant Colony Optimization

The ant colony optimization (ACO) [48] algorithm mimics the natural foraging behavior of ants when in search for food. The ACO is a swarm-based algorithm [49] that attempts to solve an optimization problem by iterating between a step where candidate solutions are constructed using a pheromone model and the update of pheromone trails using the candidate solutions, in a way that pheromone trails leading to better solutions will have their value increased.

ACO was initially used to solve graph-related problems, where artificial ants follow pheromone trails to find the shortest path between a pair of nodes in the graph. The artificial pheromone trails are represented by the variable τ_{ij}, which is associated to the arc connecting node i to node j in the graph. The decision of moving an ant k from node i to node j is based on a probability computed (2.17) using the pheromone trail τ_{ij}.

$$p_{ij}^k \propto \begin{cases} \tau_{ij} & \text{if } j \text{ is a one-step neighbor of node } i \\ 0 & \text{otherwise} \end{cases} \qquad (2.17)$$

At the beginning of the optimization process a small amount of pheromone τ_0 is assigned to all arcs in the graph. In case the ant decides to follow the arc between node i and node j, the pheromone value τ_{ij} is updated according to:

$$\tau_{ij} \leftarrow \tau_{ij} + \Delta\tau \qquad (2.18)$$

where $\Delta\tau$, in the simple ACO algorithm, is a constant value.

To avoid premature converge of the algorithm, a pheromone evaporation process is executed at each iterations of the algorithm. The evaporation process is carried out for all arcs of the graph, and the pheromone values are decreased in an exponential manner:

$$\tau_{ij} \leftarrow (1 - \rho)\tau_{ij}, \quad 0 < \rho \leq 1, \text{ for all}(i,j) \qquad (2.19)$$

The simple-ACO algorithm previously described can easily be adapted to more generic optimization problems, by considering several changes in the algorithm, like adding problem constraints to the probability (2.17), to make impossible an ant moving into an infeasible node solution. Many ACO variants were successfully tested in different optimization problems [50].

Gravitational Search Algorithm

The gravitational search algorithm (GSA) [51] is a swarm-based algorithm, where potential solutions are represented by bodies whose mass is proportional to their fitness. The search for optimal solutions is guided by Newton's gravitational law. The gravitational force between two bodies is directly proportional to the product of their masses and inversely proportional to the square of the distance between them. Considering a universe with N bodies, GSA defines the gravitational force between two generic solutions, $\mathbf{x}_i$ and $\mathbf{x}_j$, as (2.20):

$$\mathbf{F}_{ij}(t) = G(t)\frac{M_i(t) \times M_j(t)}{R_{ij}(t) + \varepsilon}\left(\mathbf{x}_j(t) - \mathbf{x}_i(t)\right) \qquad (2.20)$$

where M_i and M_j represent the mass of body i and j, respectively. $G(t)$ is the gravitational constant at moment t, which is updated at every iteration. R_{ij} is the Euclidian distance between the two bodies and ε is a small value constant.

The masses, M, are update at each iteration after evaluating the fitness of all solutions. By scanning the fitness values, the worst and best values are identified based on the optimization problem, i.e., maximization or minimization. Then, the individual fitness of each solution is scaled according to (2.21). Finally, body i mass corresponds to the rate between the scaled fitness of $\mathbf{x}_i$ and the sum of the scaled fitness for all solutions (2.22).

$$m_i(t) = \frac{\text{fitness}(\mathbf{x}_i(t)) - \text{worstFitness}(t)}{\text{bestFitness}(t) - \text{worstFitness}(t)} \quad (2.21)$$

$$M_i(t) = \frac{m_i(t)}{\displaystyle\sum_{k=1}^{N} m_k(t)} \quad (2.22)$$

The gravitational constant G in (2.20) is really not a constant since its value starts at some initial value G_0 and decreases at each iteration t of the algorithm. In (2.23), a typical decrease function for this parameter is presented, where T is the maximum number of iterations to execute the algorithm.

$$G(t) = G_0 e^{\alpha \frac{t}{T}} \quad (2.23)$$

All heuristics' algorithms have stochastic characteristics; the stochastic nature in GSA is found in the total force that acts on body i, by multiplying each force component by a random uniformly distributed number, $U(0, 1)$, in the interval 0 to 1, (2.24).

$$F_i(t) = \sum_{j=1;j\neq i}^{N} \text{Rand}_{U(0,1)} F_{ij}(t) \quad (2.24)$$

Since GSA is a swarm-based algorithm, its flow is quite similar to the flow presented for the PSO, where at each iteration particles position, which in this algorithm are called bodies' position, and velocity vectors must be adjusted. The new positions produce new potential solutions, $\mathbf{x}_i$, from trajectory's defined by the new velocity vectors:

$$v_i(t+1) = \text{Rand}_{U(0,1)} \times v_i(t) + a_i(t) \quad (2.25)$$

$$\mathbf{x}_i(t+1) = \mathbf{x}_i(t) + v_i(t+1) \quad (2.26)$$

where a_i is the acceleration of body i with mass M_i and when a force F_i is exerted on it, which can be calculated from Newton's second law of motion equation:

$$a_i(t) = \frac{F_i(t)}{M_i(t)} \quad (2.27)$$

The algorithm runs until a stopping criterion is met, which in GSA is, typically, reaching a maximum number of iterations.

Genetic Algorithm

The genetic algorithm (GA) was first proposed by Holland [52]. This optimization algorithm is based on the Darwinian principle of evolution through natural selection, where the fittest individuals will survive and the genes that make them better are inherited by their offspring.

GA operates on a population where each individual represents a potential solution ranked according to a fitness function. The individuals are defined by their chromosomes which encode a particular solution in their genes. The artificial evolutionary process starts from a random population and is based on three stages:

- Evaluation—At this stage, the algorithm computes the fitness of each individual in the population based on the potential of the solution it represents.
- Selection—The selection process applies a technique to select which individuals from the current generation are going to be the parents of the next generation. Among the most common selection techniques are:
  - *Roulette Wheel Selection* principle follows the ideas of a roulette wheel where each slot is proportional to the individual's fitness. The probability, p_i, of an individual being selected according to its fitness, f_i, is given by: $p_i = f_i / \sum_{j=1}^{n} f_j$.
  - *Linear Rank Selection* starts by sorting the N individuals of the current population based on their fitness. The best individual gets rank N and the worst gets rank 1. Based on the ranking system the selection probability is given by: $p_i = {}^{\text{Rank}_i}/_{N(N-1)}$.
  - *Tournament Selection* randomly selects, with equal probability, k individuals from the population. These individuals compete among them and the winner is the individual with the highest fitness value. The selection process is repeated n times using the entire population. The selection probability is given by: $p_i = {}^{C_{n-1}^{k-1}}/_{C_n^k}$ if $i \in [1, n - k - 1]$ and zero otherwise [53].
- Recombination or crossover is the process by which selected individuals' genes are recombined to produce new individuals or potential solutions for the next generation. Typically, two individuals are selected as parents and from an order interchange of genes two offspring are produced. In Fig. 2.11, this process is depicted.

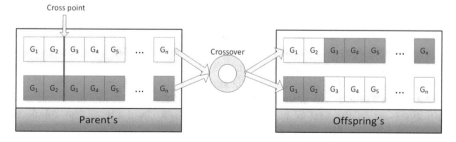

Fig. 2.11 Genetic algorithm crossover process

The presented example shows a crossover operation where a single cross point was randomly selected. The cross point defines the gene where chromosomes are broken to allow the cross connection between the segments of genes to produce the offspring.

Like in the previous described algorithms, GA has also a technique to avoid premature convergence and help solution space exploration, which is referred as *mutation* process. The mutation process is performed after *crossover* and consists in selecting several individuals from the population and setting random values to a small number of their genes. The evolutionary process is repeated until a stopping criterion is met, typically, a maximum number of generations is defined.

Algorithm 2.4 Genetic Algorithm

Initialization
$\mathbf{X}_0=[\mathbf{x}_1,\mathbf{x}_2,\ldots,\mathbf{x}_N]$; $\mathbf{F}_0=[f_1,f_2,\ldots f_N]$
For $i=1$ to N
    $\mathbf{x}_i(0) \leftarrow Rand()$ Sets random initial solutions
    $f_i \leftarrow fitness(\mathbf{x}_i(0))$ Evaluates the fitness
End For

Find solution
$t \leftarrow 0$
While *Not StoppingCriterionMet* do
    $P_t=Selection(\mathbf{X}_t, \mathbf{F}_t)$ Selection, returns parents
    $O_t=MutationCrossover(P_t)$ Performs crossover and mutation, returns offspring
    $F_O=Evaluation(O_t)$ Evaluate offspring
    $\mathbf{X}_{t+1}=NewPopulation(\mathbf{X}_t, O_t)$
    $t \leftarrow t+1$
End While
return *Best Solution*.

Nondominated Sorting Genetic Algorithm II

All heuristic algorithms presented so far are single-objective optimization algorithms. The Nondominated Sorting Genetic Algorithm II (NSGA-II) is a GA-based multi-objective algorithm developed and presented by Deb et al. in [54]. Dealing with multi-objective functions requires adding additional stages to GA. In multi-objective problems, selecting the best solutions at each generation of the algorithm is not straightforward like in single-objective problems. NSGA-II selects the best solutions based on a criterion of dominance, Fig. 2.6, and crowding distance. The *nondominated sorting* stage defines the dominance criterion, by ordering all the solutions in Pareto fronts according to their level of dominance, i.e., nondominated solutions are in the first front, in the second Pareto front are all the solutions dominated by the first front, which dominate the rest of solutions, and so forth. Following the *nondominated sorting* task is the *crowding distance* stage. Crowding

Fig. 2.12 Crowding
distance of solution X_1

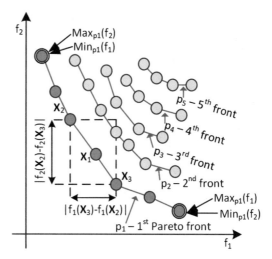

distance is a front wise value assigned to all solutions. This value is based on the
normalized Manhattan distance between the neighbor solutions which define a
hyper-box surrounding the solution of interest.

Consider the example in Fig. 2.12, where the crowding distance, CD_{X1}, is being
calculated for solution X_1 in the first Pareto front, $p1$, as:

$$CD_{X1} = \frac{|f_1(X_3) - f_1(X_2)|}{\text{Max}_{p1}(f_1) - \text{Min}_{p1}(f_1)} + \frac{|f_2(X_2) - f_2(X_3)|}{\text{Max}_{p1}(f_2) - \text{Min}_{p1}(f_2)} \qquad (2.28)$$

where $\text{Max}_{pi}(f_x)$ and $\text{Min}_{pi}(f_x)$, respectively, return the maximum and minimum
values of the objective function f_x for Pareto front level p_i.

The idea behind crowding distance is to enforce long and well-spread solutions at
each Pareto front, which ensures diversity of solutions. In order to maintain long
Pareto, the solutions at the extremes of the front will have assigned infinite crowding
distance.

The NSGA-II complete flow is presented in Algorithm 2.5, the stopping criterion
can be determined by setting a maximum number of generations or when the first
Pareto front stops improving.

The nondominated sorting with crowding distance techniques can be easily
implemented in other single-objective population-based optimization algorithms to
develop new multi-objective optimization techniques. In fact, not only population-
based optimization techniques can use Pareto front ranking and crowding distance to
create multi-objective optimization techniques, in [55] a multi-objective PSO algorithm is presented using this technique.

Algorithm 2.5 NSGA-II Pseudo Code

Initialization

$P_0=[x_1,x_2,\ldots, x_N]$ Initial random population

$Fp=Evaluation(P_0)$ Evaluate initial population

NSAG-II

$t \leftarrow 0$

While *Not Stopping Criterion Met* do

$O_t=SelectionMutationCrossover(P_t)$ Performs crossover and mutation, returns offspring

$F_O=Evaluation(O_t)$ Evaluate offspring

$N_t=Non\text{-}Dominated\text{-}Sorting(P_t \cup O_b)$ Performs non-dominated sorting, return solutions

ranked in Pareto fronts

$C_t=CrowdingDistance(N_t)$ Calculates crowding distance

$P_{t+1}=NewPopulation(N_t ,C_t)$ Using crowding distance and Pareto front rank
                                           select best solutions for new population.

$t \leftarrow t+1$

End While

return *First Pareto Front Solutions*

Multi-objective Evolutionary Algorithm Based on Decomposition (MOEA/D)

The MOEA/D [56] is a multi-objective evolutionary optimization algorithm that decomposes the multi-objective problem into N single-objective problems. There are several techniques to perform decomposition, such as Weighted Sum, Tchebycheff, and Boundary Intersection.

MOEA/D starts by defining a set of N evenly spread weight vectors, and after computing the Euclidian distances between any two vectors, create sets $B(i)$ of the T nearest neighbor vectors, where $i = 1, \ldots, N$. Next, an initial population of N individuals is created and evaluated. In case the adopted decomposition technique requires a reference point, then point z is defined based on the objective function values so far; as an example each point z component can correspond to the minimum values of each of the objective functions. Then, a neighbor set $B(i)$ is randomly selected; from the neighbor $B(i)$ two indexes are selected which are used to identify two individuals from the population. Using some crossover operation from evolutionary algorithms, a new individual is created having as parents the two previous select individuals. The new solution is then evaluated, and the reference point is updated. Also, the new solution is compared to the rest of solutions in the selected neighbor, and replaces the worst solutions in neighbor if presents better value in single-objective subproblem defined by the decomposition function. The process is repeated until a stopping criterion is reached. This algorithm presents good results for problems with more than three objectives, when compared to the NSGA-II and other multi-objective optimization algorithms using the dominance criterion, because solving the subproblems creates pressure in the Pareto front to evolve, while the dominance criterion induces low selective pressure since a better solution just needs to dominate in one of the objectives causing that most of the solutions become nondominated. Trying to overcome this problem of NSGA-II, Deb et al. in [57] presents the NSGA-III algorithm to efficiently deal with many-objective optimization problems.

In optimization literature, a distinction between multi-objective and many-objective optimization is often encountered. Although there is no consensus about the difference, [58] refers that multi-objective optimization problems with more than three objectives are many-objective problems while [59] refers to significantly more than five objectives as many-objective problems. This work adopts the definition that up to three objectives the problems are multi-objective, and beyond that number problems are many-objective.

Analog IC sizing and optimization-based approaches are essentially centered on an optimization loop, where at each iteration a potential circuit solution is evaluated in order to guide the optimization process, Fig. 2.5. The adopted technique to evaluate a solution allows to classify the optimization method. In [18] the optimization-based methods were classified into equation based, simulation based, and learning based. Each of the referred methods is described in the following sections.

2.2.2.2 Equation-Based Optimization

Equation-based evaluation techniques are based on analytic expressions deduced from the analysis of small- and large-signal equations. Knowing the equations that rule the circuit behavior allows the use of both classical and statistical methods to solve the optimization problem. In [60], the OPASYN optimization strategy is based on analytic design equations for each circuit topology; this approach was tested with several OpAmp circuits using different topologies.

In OPASYN, the parameters in the design equations were optimized using a steepest descent method. The optimization process selects as initial solutions a random set of points based on a coarse grid applied to the parameter feasible space in order to avoid getting trapped in local optimal solutions. This approach represents an evolution from knowledge-based methods but has similar flaws, like the equation model's accuracy. To identify and overcome accuracy problems, the authors validate the optimization results using SPICE simulations. Additionally, the authors refer that had to include various fitting parameters into the analytic design equations to improve model's accuracy due to the highly nonlinearity of several circuit characteristics. Another problem with this approach, like in knowledge-based approaches, is the time-consuming task related to deducing design equations for new circuits and topologies, and additionally setting the equations fitting parameters.

A similar approach was adopted in [61]. The STAIC approach adopts hierarchical circuit descriptions models that are dynamically integrated into analytical circuit equations across the different levels in the hierarchy. The multilevel parameter models are then optimized, and according to the authors "it is expected to converge to what is likely a globally optimal solution." Like in the previous OPASYN approach, this method presents similar problems, i.e., models accuracy, obtaining and maintaining circuit models.

Recently, the equation-based method was adopted to size a flexible TFT circuit. A bias-based optimization approach using LP [62] was used to size a circuit with transistors modelled using the g_m/I_d concept.

Several other methods model the relation between design parameters and circuit performances as a set of posynomial functions (2.11) and adopt a GP approach to solve the optimization problem [63–66]. Applying GP to the circuit sizing problem has the disadvantage of limiting the types of performance specifications and objectives that can be handled. As was previously referred, GP requires that the performance objective functions and constraints must be formulated in a posynomial and monomial expressions to later apply a logarithmic transformation.

Another approach using GP and posynomial functions is adopted in [67]. This work tries to address the accuracy problem of modelling the complex behavior of nanometer technology in posynomial equations, which according to the authors sometimes result in solutions that may be off by upward of 30% from the desired targets. To improve accuracy, this approach adds new error parameter terms to the equation models. These parameters are tuned during the optimization process also by GP and using SPICE simulation to validate its values.

ASTRX/OBLX [68] is a circuit sizing approach where simulation-based approach is combined with equation-based approach. The adopted optimization algorithm to find the sizing solution is the SA algorithm. ASTRX/OBLX authors defined high goals for this approach. Initial goals were reducing the preparatory effort and run-time to design new circuits; find high-quality solutions; accurate performance and device models, avoiding equation simplifications; and finally design complex high-performance circuits. Soon they start making concessions to the initial goals, like accepting a longer run-time, excluding operating ranges and manufacturability constrains. The method starts by converting the multi-objective constraint circuit sizing and optimization problem to a single-objective unconstraint optimization problem formulated as a scalar cost function (2.29), based on the objective functions and the different constrains:

$$C(\mathbf{x}) = \sum_{i=1}^{k} w_i f_i(\mathbf{x}) + \sum_{j=1}^{t} w_j g_j(\mathbf{x}) \qquad (2.29)$$

where w_i and w_j are scalar weights.

After creating the single-objective cost function, the adopted optimization algorithm must find the minimum of $C(\mathbf{x})$. During the optimization process the cost function must be evaluated, which implies calculating different circuit performances. To improve accuracy and reduce the preparatory effort, instead of using designer-supplied circuit equations, the ASTRX/OBLX use active device models' equations based on foundry models, where nonlinear devices are linearized by constructing a small-signal circuit model. Using the equation models and based on the constraints and circuit topology, the cost function equation is compiled, which allows a faster evaluation. The compiler also includes, in the cost function, all DC correctness constraints to enforce Kirchhoff's laws at each node of the circuit, by explicitly solving the DC-operating point equations. The complete evaluation of $C(\mathbf{x})$ is performed by a module using Asymptotic Waveform Evaluation [69] (AWE) simulation technique. The AWE technique is able to faster and accurately predict small-signal circuit performances using the linear performance circuit equations compiled into the cost function.

Also using SA, Meng et al. [70] adopted a two-stage optimization algorithm, which performs a global search solution based on posynomial functions that allow to roughly find an approximate solution, followed by a local search optimization process based on SA. The global search is in the form of a convex optimization problem, where circuit performance equations are model in posynomial form or any other convex formulation. The local search optimization stage adopts a SA algorithm using as initial solutions the results from the global search process. To improve solutions accuracy, which is the main drawback pointed to equation-based methods, foundry device models are adopted during local search, and circuit performances are evaluated by a circuit simulator.

Swarm intelligence techniques are also used to size analog circuits in equation-based methods. In a recent work the backtracking search technique was combined with the Ant Colony Optimization (ACO) algorithm, the resulting algorithm was named BA-ACO [71], and was tested using an Op-Amp and OTA circuits. The use of known circuit topologies helped deriving the equations used during the solutions evaluation process. The backtracking technique avoids the ACO algorithm getting trapped in local optimum. The idea behind this technique is returning to previous pheromone levels on the optimal path if the optimal values do not change for some time. The decrease in pheromone levels encourages ants to explore new paths, which may result in better solutions discovery.

In [72], the authors presented an equation-based circuit sizing approach using the PSO algorithm, and a multi-objective PSO (MOPSO) approach was also tested. The MOPSO implements the dominance and crowding distance criteria using an archive where nondominated solutions are stored during the optimization process. At the end of the execution of the algorithm, the solutions in the archive provide the Pareto front. The MOPSO algorithm was tested in one example circuit using SPICE to validate results. The standard single-objective PSO was tested in another circuit example where Agilent Advanced Design System (ADS) simulator was used for validation purposes. Again, the concerns are clear about accuracy when using equation-based methods to evaluate solutions. A modified PSO algorithm was presented in [73] to implement an automatic analog circuit sizing approach. The novelty of the modified PSO algorithm is based on the manipulation of coefficients c_1 and c_2 in (2.16); also a new random coefficient factor multiplies the inertia term. By adjusting those coefficients at each iteration, the algorithm can change from one iteration to another between exploration and exploitation phases and vice versa. This technique helped improving the convergence of the algorithm to the final solution when compared to a regular PSO algorithm.

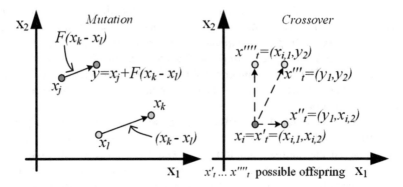

Fig. 2.13 Differential Evolution mutation and crossover

In [74] the Pareto dominance and crowding distance techniques are added to the differential evolution (DE) algorithm to create a multi-objective optimization algorithm, named MODE. The DE algorithm is inspired by the GA.

To create the offspring for the next generation, the DE algorithm for each individual $\mathbf{x_i} = [x_{i, 1} \ldots x_{i, p} \ldots x_{i, n}]^T \in \mathbb{R}^n$ in the population randomly selects three more different individuals, $\mathbf{x_j}$, $\mathbf{x_k}$, and $\mathbf{x_l}$, with $i \neq j \neq k \neq l$. Using these four individuals, the components y_p of a new individual $\mathbf{y} = [y_1 \ldots y_p \ldots y_n]^T \in \mathbb{R}^n$ are set according to $y_p = x_{j,p} + F(x_{k,p} - x_{l,p})$, where the parameter F is a real constant value named differential weight. Next, a crossover is performed using $\mathbf{y}$ and $\mathbf{x_i}$, Fig. 2.13. The crossover generates a new potential offspring $\mathbf{x_t} = [x_{t, 1} \ldots x_{t, p} \ldots x_{t, n}]^T \in \mathbb{R}^n$ by setting every component $x_{t, p}$ using a uniformly distributed random number $\mathbf{r} \sim U$ (0, 1) and according to the condition $\mathbf{r} < \mathbf{CR}$, where $\mathbf{CR} \in [0, 1]$ is a crossover parameter rate, and if the condition is true then $x_{t, p} = y_p$ otherwise $x_{t, p} = x_{i, p}$. After completely defining the new offspring $\mathbf{x_t}$, the solutions must be evaluated and if its fitness is better than the fitness of $\mathbf{x_i}$, then the new offspring replaces the individual $\mathbf{x_i}$ in the population. The MODE algorithm instead of comparing the fitness of both individuals, adds the new elements to the population and, like the NSGA-II, implements the nondominance sorting criteria with crowding distance to select the next generation individuals. The authors compared the MODE algorithm with the NSGA-II and concluded that MODE was faster and produced a better diversity of solutions.

Equation-based methods are able to fast estimate the circuit performance values, making them extremely suited to derive first-cut designs. The main drawback is the costly development of the equations and its reusability. Additionally, the approximations introduced in the equations yield low accuracy results, especially in the new nanometer scale technologies.

2.2.2.3 Simulation-Based Optimization

The high availability of fast computer resources makes simulation-based methods the most common approach found in recent works and also in commercial EDA tools. This method uses an electrical simulator, e.g., SPICE, to estimate the circuit performances with great accuracy; it is also possible to perform variability analysis and parametric yield estimation of potential sdolutions using MC simulations.

Simulation-based methods' usually implement stochastic optimization kernels, such as evolutionary computation algorithms, simulated annealing. The adoption of meta-heuristic techniques, like the genetic algorithm, allows finding optimal satisfactory solutions in a reasonable time.

An example of this approach is the AIDA-C [12, 75] analog sizing tool, which implements a multi-constraint, multi-objective genetic algorithm optimization kernel based on the NSGA-II algorithm. The complete AIDA-C framework will be described in detail in Chap. 5.

The main bottleneck in simulation-based algorithms is the evaluation stage, where each potential solution fitness is computed based on the circuit performances that are calculated by means of an electrical simulator. Depending on the set of circuit performance measures required to evaluate a solution, simulations can take from hundreds of milliseconds to several minutes. Additionally, in commercial electrical simulators before performing a simulation, a time-consuming step for license validation is required. This is why several works reduce the total population evaluation time by sharing the simulation load among multiple computers using parallelization methods [45, 76, 77]. In [76], Santos-Tavares et al. adopt a GA optimization kernel, where the population evaluation is performed in a decentralized manner using several available computers to simulate each solution. The GA is centralized in a master computer which distributes one or more solutions to be evaluated among several slave computers using standard Message Passing Interface (MPI). After performing the required simulations, in an open-source NGSPICE simulator, the slave computers return evaluation results to the master. This approach was tested to size an OpAmp circuit where the design variables were coded in a 992 bits long chromosome. The chromosome genes include variables such as drain current, channel width, and length for each transistor, three op-amp compensation capacitors, and two satellite compensation capacitors. The adopted distributed evaluation technique was able to achieve a speed up in simulation time up to 19 times when compared with single machine evaluation method.

In [77] the MAELSTROM tool is presented; this tool adopts the basic formulation from the ASTRX/OBLX approach presented in the equation-based section. MAELSTROM replaces the SA optimization kernel adopted in ASTRX/OBLX by a combined GA and SA optimization kernel, named parallel re-combinative simulated annealing (PRSA). Also, the AWE simulation technique was replaced by Spectre Cadence© electrical simulator. The adopted PRSA technique was able to overcome the difficulty of parallelizing typical SA algorithm.

The PRSA technique distributes a large initial number of potential solutions among different PRSA-nodes CPUs. At each of these PRSA-nodes a standard SA optimization process runs for a relatively small number of iterations. At each iteration, potential solutions are synchronized from one PRSA-node to a randomly selected set of other PRSA-nodes. The shared solutions can produce new potential solutions by perturbation of the design variables or using a selection and recombination process like in GA, where pairs of parent solutions are selected and finally recombined into a single offspring solution. In addition to the distributed optimization kernel, the simulation-based evaluation process of potential solutions is also distributed. The distributed evaluation process is coordinated by an evaluation master node that receives evaluation requests from every PRSA-node, and based on the type of simulation analyses decides to which slave-node sends the simulation requests. After collecting the results from the different simulations performed in the slave-nodes, the master-node returns the complete evaluated solution to the corresponding PRSA-node.

Phelps et al. in [45] presented the ANACONDA sizing tool, which was also based on the ASTRX/OBLX approach, adopting a SPS optimization kernel. ANACONDA also implements parallelism to improve execution time performance. The parallel optimization process starts from a large population of potential solutions sorted according to a cost function, and then randomly subsets are selected from this population; the random selection process is biased towards the best solutions. These subsets are sent to different computers where the SPS optimization process is executed. Next the solutions from each SPS process are added to the initial population, again the solutions are sorted, and the worst solutions are eliminated to maintain the population size. This process is repeated until no more improvements can be found in any element of the population. ASTRX/OBLX and all approaches based on it convert the circuit sizing problem, which is in its essence a multi-objective and multi-constraint problem, into a single-objective problem where a scalar weighted cost function is created based on the different objectives and constraints. This type of approach requires specifying the weight values before starting the optimization process, which, unless the designer has some knowledge about the circuit, may not be an easy task to accomplish.

In [78], Gupta et al. adopted a swarm intelligence ACO algorithm to address the analog circuit sizing task. The HSIPCE circuit simulator was used to compute circuit performances of potential solutions from ACO algorithm. The objective or cost function implemented in the optimization algorithm corresponds to the squared root of sums of the normalized error between the simulated performance and the desired performance:

$$ f = \sqrt{\sum \left(\frac{p_i^d - p_i^s}{p_i^d} \right)^2} \tag{2.30} $$

where p_i^d is performance i desired or target value and p_i^s is the simulated value for the same performance.

The adoption of cost functions like (2.30), for an analog IC sizing problem, results in feasible solutions that only comply with the target performances and where no particular performance is optimized, since as soon the cost function becomes zero no further optimization is possible. The authors compared several runs of ACO sizing results with a GA approach, using the same cost function and design variables, and with a manual design solution. The paper does not provide many details about important parameters for the optimization algorithms, such as the size of the population of the GA. The presented results showed that, for most of the circuits sized, the best GA solutions were better than the ACO solutions, but the comparison between different runs show less consistent results for the GA approach. Results also present better execution times for the ACO algorithm, allowing the authors to conclude that on average the ACO approach attains better results and that the ACO algorithm is an acceptable alternative to the GA approach. Another interesting conclusion from the results revealed that the ACO solutions were able to achieve better performance values than the manual design circuits. A similar approach was followed by Benhala et al., in [79], comparing the GA with the ACO algorithm. The adopted cost function was also an error function between the simulated and required performances. The comparison between the two algorithms revealed that the GA was faster and requires less algorithm parameters than ACO, which does not correspond to the conclusions in Gupta et al. work, in [78]. Nevertheless, in both works it was concluded that ACO is able to provide on average better solutions than the GA.

In [80] a hybrid approach that combines two swarm-based techniques is presented; a swarm algorithm based on the GSA is combined with PSO. The resulting hybrid algorithm, named Advanced GSA_PSO, was tested first using some known benchmark functions and next applied to the circuit sizing problem, where circuit performances were computed by an HSPICE electrical simulator. Advanced GSA_PSO is based on another hybrid PSOGSA approach where the velocity expression (2.25) becomes similar to the PSO velocity:

$$v_i(t + 1) = \text{Rand}_{U(0,1)} \times v_i(t) + c_1 a_i(t) R_1 + c_2 (g - x_i(t)) R_2 \qquad (2.31)$$

The Advanced GSA_PSO hybrid approach adopts the social component from the PSO algorithm and adds a technique named *shrinking circles* at each iteration of the algorithm. The shrinking circles technique identifies the position of the best solution so far, which for the GSA is the largest mass body. Afterwards, a circle is drawn around this point with a radius that shrinks during the algorithm execution. Inside this circle a growing number of randomly generated new bodies or potential solutions are created. The new bodies are evaluated and in case one of those random bodies has better fitness than the initial best solution, the new highest fitness solution replaces the old one in the population.

The shrinking circles technique, according to the authors, helps the GSA algorithm to find better solutions, since premature convergence is avoided. However, the shrinking circles technique requires the definition of additional parameters which may imply an additional setup time.

2.2.2.4   Model-Based Optimization

Learning or model-based method's main advantage is the fast evaluation of potential solution during the optimization process. This approach is based on models inferred from electrical simulations. The huge amount of data and pre-processing efforts needed to train the models is one of the main drawbacks of this approach. Another difficulty is the reusability of the models, since a simple change in the circuit makes the models unfitted for purpose.

Surrogate models are one of the learning techniques used to model cost and constraint functions [81, 82]. In this work, the surrogate model term is loosely used to describe a technique capable of modelling complex system with a lower computational cost and with an acceptable accuracy. Mainly, surrogate models in automatic analog IC sizing solutions aim replacing the electrical simulator as the circuit performance evaluator by a more time efficient evaluation technique.

Surrogate model's construction is based on the optimization of several model parameters, typically according to the minimization of a representation error function, such as the root mean squared error. The phase during which these parameter values are set is called the training phase. Surrogate model's training requires vectors of data that include samples of the input variables and the corresponding output. Sampling is a very important part of the surrogate model construction. The question of how many samples are enough to obtain a model that effectively represents the real design variable space does not have an easy answer. The number of samples depends on the dimensionality of the design space, i.e., the number of variables and the range of the variables. To illustrate this problem, consider different dimensional design variable spaces, from one dimension up to three dimensions, and where the range of variables goes from zero to fifteen for each dimension. Also, consider that the space is divided in intervals of size five at each dimension and ten random samples are generated according to the variables ranges. In Fig. 2.14, the samples are plotted at the different dimensional spaces and the slots are identified. In the three-dimensional space, Fig. 2.14c, the projections of the samples are marked to easily identify the relative sample position.

At the one-dimensional space, Fig. 2.14a, each of the three slots has several samples in it. When the space dimensionality grows to two, Fig. 2.14b, nine slots of size 5×5 appear and two of them have no samples. If again another dimension is added to the variable space, Fig. 2.14c, now 27 slots exist inside the variable range, and the samples start to become sparse in this three-dimensional space. So, each time a design variable is added, and the design variable space grows, more samples are required to evenly represent the design variable space and train the surrogate models. The exponential growth in the number of design variable samples needed to model a function was coined by Bellman as the "curse of dimensionality" in [83]. Also, a higher number of samples is required, if the variable ranges increase.

The samples used in surrogate model's training phase are often obtained by expensive circuit simulations, so it is very important keeping the number of samples to a minimum. Several techniques were developed and adopted to reduce design

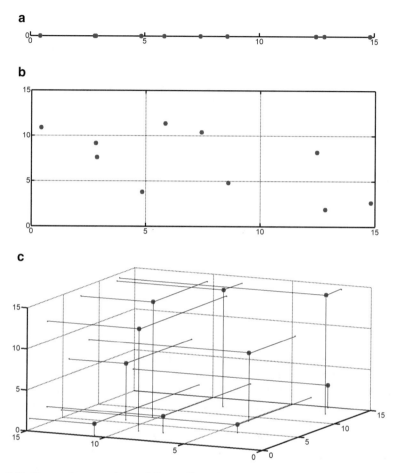

Fig. 2.14 Ten random samples at different dimensionality spaces. (**a**) One-dimensional space. (**b**) Two-dimensional space. (**c**) Three-dimensional space

variable space dimensionality, and, also, to evenly sample the variable space. A simple technique to reduce space dimensionality is performing a sensitivity analysis. Using this differential technique, it is possible to understand how changes in each design variable reflect on the output. In sensitivity analysis, design variables with low impact on the output are discarded and only variable that produces significant changes at the output are considered for training purposes.

Principal component analysis (PCA) [84] is another common technique adopted to achieve space dimensionality reduction. In PCA the covariance matrix of N samples, $\mathbf{x_i} = [x_1 \ldots x_d]^T \in \mathbb{R}^d$, is computed:

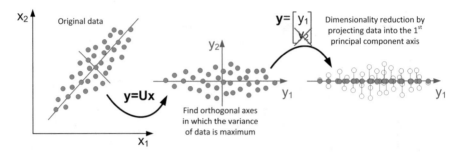

Fig. 2.15 Principal component analysis

$$C = \frac{1}{N} \sum_{i=1}^{N} (\mathbf{x}_i - \boldsymbol{\mu})(\mathbf{x}_i - \boldsymbol{\mu})^T \qquad (2.32)$$

with

$$\boldsymbol{\mu} = \frac{1}{N} \sum_{i=1}^{N} \mathbf{x}_i \qquad (2.33)$$

Next the eigenvectors and eigenvalues of the covariance matrix $\boldsymbol{C}$ are calculated. Denoting $\mathbf{U}$ as a matrix with columns defined by the eigenvectors sorted descending according the eigenvalues, it is possible to calculate the transformation $\mathbf{y} = \mathbf{U}\mathbf{x}$. The eigenvectors from $\boldsymbol{C}$ define an orthogonal basis with axes oriented according to the directions of larger variances of the samples, and by considering only the first p rows of $\mathbf{y}$, with $p < d$, results in a projection of sample $\mathbf{x}$ into the main or principal p axes; thus some dimensions of lower variance are eliminated, Fig. 2.15.

In order to better mimic functions, surrogate model's training process must receive data samples that are representative of the design variable space. Several sampling techniques are used to determine the locations of sampled data points and reduce the number of required samples [85], such as Quasi-Monte Carlo (QMC), Latin Hypercube Sampling (LHS), Uniform Design (UD), and Orthogonal Array Design (OAD).

QMC sampling implements a Low Discrepancy Sequence (LDS) generator. LDS are specifically designed to obtain samples as uniformly as possible from the design variable space. To measure the deviation of samples from a uniform distribution, consider an $\mathbf{S}$ d-dimensional unit rectangular space, where N samples $\mathbf{x}_i = [x_1 \ldots x_d]^T \in \mathbb{R}^d$, $\mathbf{i} = 1, \ldots, N$, were produced. Consider also a smaller $R(\boldsymbol{l})$ d-dimensional rectangular space with volume $m(R(\boldsymbol{l}))$, such that $R(\boldsymbol{l}) \subset \mathbf{S}$, where $\boldsymbol{l} = (l_1, \ldots, l_d)$ defines the length of the sides of the rectangle, Fig. 2.16. Then the local discrepancy is defined as (2.34), [86].

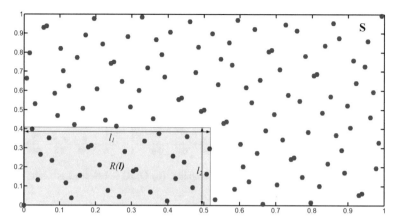

Fig. 2.16 Local discrepancy, $d(I)$, using a Sobol LDS with $N = 128$

$$d(I) = \frac{N_{R(I)}}{N} - m(R(I)) \tag{2.34}$$

where $N_{R(I)}$ is the number of samples from the N samples that exist inside $R(I)$.

Based on $d(I)$, the star discrepancy can be defined as:

$$D_N^* = \sup_{R(I) \,\in\, \mathbf{S}} |d(I)| \tag{2.35}$$

Finally, an LDS is a sequence that satisfies the following condition:

$$D_N^* \leq C(d) \frac{(\ln N)^d}{N} \tag{2.36}$$

where $C(d)$ is a constant whose values depend upon the dimension of space and the sequence, but not on the number of samples.

QMC samples can be generated using several LDS methods, such as Halton or Sobol sequences [87], Fig. 2.17. QMC main drawback is related with its application to problems with high-dimensional spaces, where the sequence generation can be a time-consuming task. Also, some LDS generators can produce sequences not completely random, at those high-dimensional spaces, revealing correlations among samples [88].

LHS is a sampling technique widely adopted, mainly by its simplicity and understanding of how the sampling generation is achieved. Consider a range $[0, 1]$ divided in N equal sized intervals. From each interval a random point is selected, defining a sequence of N random points $S_1 = \{x_1, \ldots, x_N\}$, $x_i \in \mathbb{R}^1$. Using the same approach, another sequence $S_2 = \{y_1, \ldots, y_N\}$, $y_i \in \mathbb{R}^1$ is created. Then, it is possible to randomly pair elements from both sequences to create a bidimensional random sample sequence, $S_{12} = \{(x_5, y_1), (x_2, y_7), \ldots, (x_{N-b}, y_{N-a})\}$. Adopting this idea, it

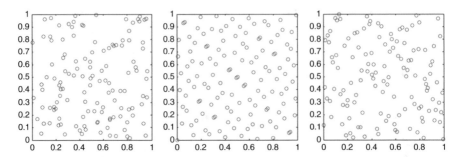

Fig. 2.17 Comparison between (**a**) random sampling, (**b**) QMC Sobol sequence, and (**c**) Latin hypercube sampling for 128 samples

is possible to keep pairing any number of one-dimensional random sequences to create a random sample sequence for any d-dimensional space.

Different modelling techniques are used to model the nonlinear behavior of analog circuits. In [89], LHS is adopted to train a modelling technique known as Kriging Model [90]. The general idea of Kriging modelling is that sampled points near the point being evaluated must have a higher weight or contribution in the evaluation than sample points located further away. The model includes the spatial relation among the different points in the form of a covariance. The example considered was quite simple, where only two circuit performances were evaluated, forcing to create two distinct Kriging models. Also, only four design variables were used to size the circuit, which justifies that only 24 samples were enough to train the model. LHS was also adopted in [91] to train three different modelling techniques. The different models were included in an optimization process and the results were compared with another optimization run using a simulation-based approach. In the optimization process three circuit performance measures and four design variables were considered. The models were created using Kriging, an interpolation method using Radial basis function [92], in its linear form, $\widetilde{f}(\mathbf{x}) = \sum_{i=1}^{N} w_i \phi(\|\mathbf{x} - \mathbf{x}_i\|), \mathbf{x} \in \mathbb{R}^d$ where $\phi(r) = r$, and the last modelling technique adopted implements a rational function model [93] as a ratio between two second-order polynomials, $\widetilde{f}(\mathbf{x}) = P(\mathbf{x})/Q(\mathbf{x})$, where $P(\mathbf{x})$ and $Q(\mathbf{x})$ are polynomial in the form $p(\mathbf{x}) = w_0 + \sum_{i=1}^{N} w_i x_i + \sum_{i=1}^{N} w_{ii} x_i^2 + \sum_{i=1}^{N} \sum_{j=1}^{N} w_{ij} x_i x_j$. As expected, the adoption of models allows evaluating the solutions much faster than the electrical simulator, thus reducing the optimization time. The authors reported the time for training the models and the number of samples used, but do not reveal the samples generation and setup time.

Support Vector Machine (SVM), a machine learning technique, was adopted by Bernardinis et al. in [94], to model analog circuit performance spaces. This work intends modelling the behavior of a circuit at higher levels of abstraction based on simulated performance values. To achieve the goal of modelling circuit behaviour, this work adopts SVM to define approximating performances relations. The SVM classifier defines the boundaries of feasibility regions considering performances values. Two methods were used to train the classifier: a two-class SVM—this method requires the existence of points in the training set with two different outcomes; since all simulated solutions are feasible, a set of infeasible circuits was generated based on density functions; the second training method was the one-class SVM—in this method were considered feasible all simulated solutions and are assumed infeasible any solutions outside the training set. On tests, the one-class SVM reveal better modelling accuracy. The SVM models developed in this work aim to answer questions at higher levels of abstraction such as: given a feasible relation among performances $P(p_1, p_2, p_3, p_4, p_5) = 1$, where p_i are different performances, is it possible to define a feasible relation $P(p_1, p_2, p_3) = 1$ for p_4 and p_5 with some specific value, or having two performance relations is it possible combining the two to obtain a feasible solution.

Barros et al., in [95], also uses SVM to model the performance space and identify the feasible design space regions. In this work, SVM models are used at circuit synthesis level to help the adopted GA optimization kernel identifying promising regions in the design variable space, and, also, to model the cost function and avoid expensive simulations. The feasibility classification and performance regression SVM models were trained using 14,000 samples, where 20% of them were used as validation set for both models. Tests were carried out using a two-stage amplifier with results showing significant gains in time efficiency of the overall sizing and optimization process.

In [96], Alpaydin et al. presented a circuit synthesis solution based on a single-objective optimization process combining GA with SA. The circuit ac performances of potential solutions were computed using neural-fuzzy models trained from SPICE simulations. The neural-fuzzy models adopted employ a three-layer neural-fuzzy network to estimate circuit performances and avoid expensive electrical simulations. At the first layer or fuzzification layer of the network a Gaussian function was adopted as membership function. The second layer uses the product operation rule of fuzzy implication [97], and the output layer uses as defuzzifier the center-of-gravity method [97]. The training process was performed by an evolutionary optimization procedure using data samples simulated in HSPICE according to LHS.

Model-based approaches have limited application when design variable space exploration is required to size new circuits, since for creating the models training data is required from working circuit solutions. Additionally, the computational burden of developing different models, one for each circuit performance, limits its use to applications with a small number of performances or developing a classification model to avoid simulation of points that are potentially infeasible.

2.3   Circuit Design and Performance Parameters

As stated before, circuit sizing task is where circuit design variables or parameter values are set. A yield-aware sizing solution must obtain a combination for those parameters taking under consideration the effects of variability. Variability parameters, design variable parameters, and environment parameters which affect circuit performances are defined as follows:

- *Design parameters* are circuit parameter variables that designers are able to tune in order to adjust circuit performances according to the required specifications. Design parameters assume nominal values, typically inside a range, either defined by the selected manufacturing process technology or by specifications defined at higher hierarchical levels in the HAD flow. Examples of this type of parameters or design variables are nominal transistors' gate width (W) and length (L) selected from a range of allowable values defined by the adopted technology, $W \in [W_{min}, W_{max}]$. Design parameters are represented as a vector $\mathbf{x}_d = [x_{d,\ 1} \ldots x_{d,\ n}]^T \in \mathbb{R}^n$, which define the n-dimensional design variable space D. In nominal circuit sizing, the values in this vector are set such that the desired circuit performances are achieved.

- *Statistical parameters* reflect how random variations, due to nonideal manufacturing processes and materials, impact on the different devices present in the circuit and consequently circuit performances. As was previously referred, these random variations are modelled as probability density functions of the statistical parameters, which add the variability effects to a nominal parameter value. Several of the nominal parameter values result from design parameters values, like the transistors' gate width to which is applied a variation component according the probability density function of the gate width statistical parameter. Other types of statistical parameters are intrinsic model parameters of the different electronic devices in the circuit, such as threshold voltage and oxide thickness. Statistical parameters can be represented as vector $\mathbf{x}_s = [x_{s,\ 1} \ldots x_{s,\ m}]^T \in \mathbb{R}^m$, which reflects process variations that define a process parameter space S with a distribution pdf($\mathbf{x}_s$).

- *Environment parameters* are, typically, range parameters whose values are derived from several operating conditions, which is why they are also referred as operating parameters. Examples for this type of parameters are supply voltage and temperature. Although an environment parameter might have associated a statistical distribution, the circuit solution must fulfill all the desired specifications for the operating condition range regardless of the distribution. For instance, a supply voltage environment parameter with values in a range of 3–8 V requires that the circuit solution must comply with the required specifications when operating in that interval of values for the supply voltage. However, the supply voltage might be originated in a voltage source with a voltage ruled by some Gaussian distribution where the mean is the nominal voltage, for instance 5 V, and a 6σ variation of 1.5 V, resulting in a most likely variation voltage in the range of 3.5–6.5 V. As can be deduced from the presented example, environment parameters are usually derived from the circuit specifications, but since they are numerical parameters that affect circuit performances, they must be considered as parameters to incorporate in the circuit validation step. Like for the other type of parameters, environment parameters can be expressed as a vector $\mathbf{x}_e = [x_{e,1} \ldots x_{e,k}]^T \in \mathbb{R}^k$.

2.3.1   Feasibility Regions

Based on these three types of parameters it is possible to define a vector including all numerical values that affect the output performances of a circuit:

$$\mathbf{x} = [x_{d,1} \ldots x_{d,n}, x_{s,1} \ldots x_{s,m}, x_{e,1} \ldots x_{e,k}]^T = [x_1 \ldots x_{n+m+k}]^T \in \mathbb{R}^{n+m+k} \quad (2.37)$$

From the definition of the different parameters it is possible to conclude that most of them are limited to a specific range of values. The limits of those ranges define the boundaries for a feasibility design parameter space regardless of the circuit performance specifications, since design parameter values outside those boundaries are not allowed, typically, due to some technology limitation. The combination of the different feasibility spaces from each design parameter results on different types of feasibility regions. Range parameters, like environment parameters or some design parameters such as transistors gate width/length, create rectangular box regions resulting from the intersection of the different conditions and planes formed by the limits of the range. Having two range parameters $R_1 = \{x_1 | x_{1,L} \leq x_1 \leq x_{1,U}\}$ and $R_2 = \{x_2 | x_{2,L} \leq x_2 \leq x_{2,U}\}$ it is possible to represent the intersection of the different inequalities, $x_{1,L} \leq x_1 \cap x_1 \leq x_{1,U} \cap x_{2,L} \leq x_2 \cap x_2 \leq x_{2,U}$, like the rectangular region in Fig. 2.18 as the feasibility region.

Fig. 2.18 Rectangular feasibility box region

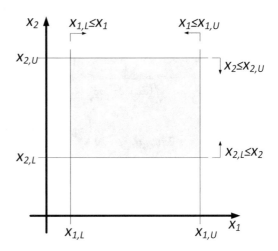

Another common feasibility region is ellipsoids, which are mainly defined by Gaussian statistical parameters $(\mathbf{x}_s)$ intersected by level planes.

Considering the probability density function $\text{pdf}(\mathbf{x}_s)$, (2.39), as an m-dimensional Gaussian with mean vector $\mathbf{x}_{s_0} = [x_{s_0,1} \ldots x_{s_0,m}]^T \in \mathbb{R}^m$ and a covariance matrix $\mathbf{C}^{m \times m}$ (2.40):

$$\mathbf{x}_s \tilde{N}_m(\mathbf{x}_{s_0}, \mathbf{C}) \tag{2.38}$$

$$\text{pdf}(\mathbf{x}_s) = \frac{1}{\sqrt{(2\pi)^m |\mathbf{C}|}} e^{\left(-\frac{1}{2}(\mathbf{x}_s - \mathbf{x}_{s_0})^T \mathbf{C}^{-1}(\mathbf{x}_s - \mathbf{x}_{s_0})\right)} \tag{2.39}$$

with

$$\mathbf{C} = \begin{bmatrix} \sigma_{x_{s,1}}^2 & \sigma_{x_{s,1}}\sigma_{x_{s,2}}\rho_{x_{s,1,2}} & \cdots & \sigma_{x_{s,1}}\sigma_{x_{s,m}}\rho_{x_{s,1,m}} \\ \sigma_{x_{s,2}}\sigma_{x_{s,1}}\rho_{x_{s,2,1}} & \sigma_{x_{s,2}}^2 & \ddots & \vdots \\ \vdots & \ddots & \ddots & \sigma_{x_{s,m-1}}\sigma_{x_{s,m}}\rho_{x_{s,m-1,m}} \\ \sigma_{x_{s,m}}\sigma_{x_{s,1}}\rho_{x_{s,m,1}} & \cdots & \sigma_{x_{s,m}}\sigma_{x_{s,m-1}}\rho_{x_{s,m,m-1}} & \sigma_{x_{s,m}}^2 \end{bmatrix} \tag{2.40}$$

where $\sigma_{x_{s,i}}$ is the standard deviation of $x_{s,\,i}$ and $-1 \leq \rho_{x_{s,i,j}} \leq 1$ defines the correlation between $x_{s,\,i}$ and $x_{s,\,j}$.

Multivariate Gaussians' are easily understood by the shape of its isocontours or curve-levels. An isocontour for a function $f : \mathbb{R}^n \to \mathbb{R}$ is defined as:

$$\{\mathbf{x} \in \mathbb{R}^n, c \in \mathbb{R} : f(\mathbf{x}) = c\} \tag{2.41}$$

To deduce isocontour shapes for Gaussian curves consider, (2.42)–(2.48), without any loss of generality and for simplicity, two uncorrelated Gaussian variables:

$$m = 2; \quad \mathbf{x}_s = [x_{s,1}, x_{s,2}]^T; \quad \text{with mean } [x_{s_0,1}, x_{s_0,2}]^T \tag{2.42}$$

$$\text{pdf}(\mathbf{x}_s) = c \Rightarrow c$$

$$= \frac{1}{\sqrt{(2\pi)^2 \sigma_{x_{s,1}}^2 \sigma_{x_{s,2}}^2}} e^{\left(-\frac{1}{2\sigma_{x_{s,1}}^2}(x_{s,1} - x_{s_0,1})^2 - \frac{1}{2\sigma_{x_{s,2}}^2}(x_{s,2} - x_{s_0,2})^2\right)} \tag{2.43}$$

$$2\pi\sigma_{x_{s,2}}\sigma_{x_{s,1}} c = e^{\left(-\frac{1}{2\sigma_{x_{s,1}}^2}(x_{s,1} - x_{s_0,1})^2 - \frac{1}{2\sigma_{x_{s,2}}^2}(x_{s,2} - x_{s_0,2})^2\right)} \tag{2.44}$$

$$\log\left(\frac{1}{2\pi\sigma_{x_{s,2}}\sigma_{x_{s,1}} c}\right) = \frac{1}{2\sigma_{x_{s,1}}^2}(x_{s,1} - x_{s_0,1})^2 + \frac{1}{2\sigma_{x_{s,2}}^2}(x_{s,2} - x_{s_0,2})^2 \tag{2.45}$$

$$1 = \frac{(x_{s,1} - x_{s_0,1})^2}{2\sigma_{x_{s,1}}^2 \log\left(\frac{1}{2\pi\sigma_{x_{s,2}}\sigma_{x_{s,1}}c}\right)} + \frac{(x_{s,2} - x_{s_0,2})^2}{2\sigma_{x_{s,2}}^2 \log\left(\frac{1}{2\pi\sigma_{x_{s,2}}\sigma_{x_{s,1}}c}\right)} \tag{2.46}$$

defining

$$a = \sqrt{2\sigma_{x_{s,1}}^2 \log\left(\frac{1}{2\pi\sigma_{x_{s,2}}\sigma_{x_{s,1}}c}\right)}$$

and

$$b = \sqrt{2\sigma_{x_{s,2}}^2 \log\left(\frac{1}{2\pi\sigma_{x_{s,2}}\sigma_{x_{s,1}}c}\right)} \tag{2.47}$$

it follows that

$$1 = \left(\frac{x_{s,1} - x_{s_0,1}}{a}\right)^2 + \left(\frac{x_{s,2} - x_{s_0,2}}{b}\right)^2 \tag{2.48}$$

Expression (2.48) is the equation of an ellipse aligned with the main axes, due to the uncorrelation between variables. The ellipse has center in $(x_{s_0,1}, x_{s_0,2})$, and the axis of the ellipse according to $x_{s,\,1}$ axis has length $2a$ and according to $x_{s,\,2}$ has length $2b$. In Fig. 2.19 an example of a bivariate Gaussian distribution is presented; in Fig. 2.20 several isocontours, in the form of different color tone ellipses, are presented.

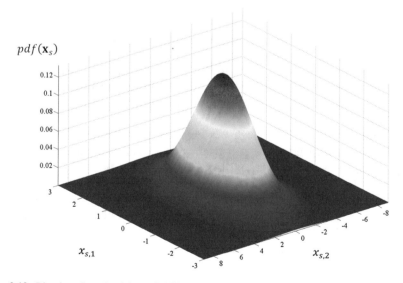

Fig. 2.19 Bivariate Gaussian join probability density function

Fig. 2.20 Isocontours ellipses of bivariate Gaussian

Isocontours for multivariate Gaussian distributions with correlated variables are ellipsoids whose orientation is related to the correlation between the different variables, and are expressed as:

$$(\mathbf{x}_s - \mathbf{x}_{s_0})^T \mathbf{C}^{-1} (\mathbf{x}_s - \mathbf{x}_{s_0}) = \beta^2(\mathbf{x}_s) \qquad (2.49)$$

where β defines the scale of the ellipsoid. For $\beta = 0 \Rightarrow \mathbf{x}_s = \mathbf{x}_{s_0}$, the pdf($\mathbf{x}_s$) assumes its maximum value $\frac{1}{\sqrt{(2\pi)^m |\mathbf{C}|}}$.

As the value of β increase, the ellipsoids scale increase while the pdf($\mathbf{x}_s$) function value exponentially decreases.

An example for an ellipsoid feasibility region, which is defined as the region inside an isocontour that can be obtained from the generic expression $E = \{\mathbf{x} = [x_1, x_2]^T | (\mathbf{x} - \mathbf{x_0})^T \mathbf{A}(\mathbf{x} - \mathbf{x_0}) \leq B\}$ with $\mathbf{A}$ positive definite matrix, is presented in Fig. 2.21.

A new type of feasibility region is obtained by the intersection of different parameter planes which define polytope region. An example of a n-dimensional polytope regions is $P = \{\mathbf{x} = [x_1, x_2]^T | \mathbf{A}\mathbf{x} \leq \boldsymbol{b}\}$, where $\mathbf{A} \in \mathbb{R}^{n \times 2}, \boldsymbol{b} \in \mathbb{R}^n$, Fig. 2.22.

Finally for nonlinear parameters, the intersection of the different nonlinear function boundaries results in a nonlinear region of feasibility, like in Fig. 2.23, $N = \{\mathbf{x} = [x_1, x_2]^T | \boldsymbol{\Phi}(\mathbf{x}) \geq \mathbf{0}\}$, where $\boldsymbol{\Phi}(\cdot)$ is like a vector of n nonlinear functions,

i.e., $\boldsymbol{\Phi}(\mathbf{x}) = \begin{bmatrix} f_1(\mathbf{x}) \\ \vdots \\ f_n(\mathbf{x}) \end{bmatrix}$, where $f_i(\mathbf{x})$, with $i = 1, \ldots, n$, are nonlinear functions and $\mathbf{0}$ is

a vector of n zeros.

Fig. 2.21 Ellipsoid
feasibility region

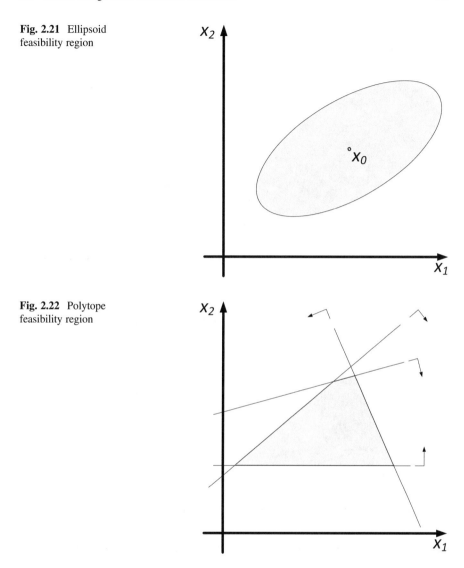

Fig. 2.22 Polytope
feasibility region

2.3.2   *Circuit Design and Performance Parameter Space Relation*

Performance features are important values that allow characterizing a circuit according to some key features, e.g., DC gain, phase margin, and power consumption. Performance values are obtained, for a specific set of circuit parameters or solution, either by simulation in an electrical simulator or by some circuit model that relates circuit design parameters with performances. So, a circuit behaves like a

Fig. 2.23 Nonlinear
feasibility region

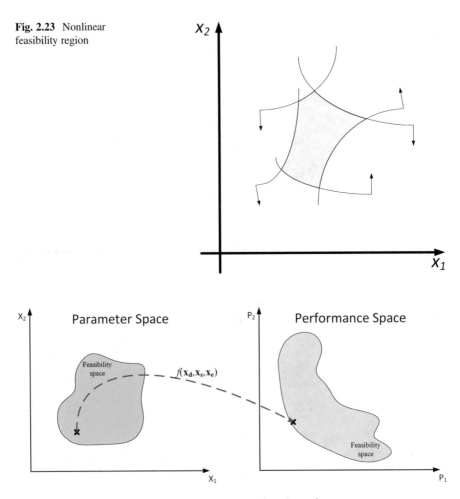

Fig. 2.24 A circuit maps points in the parameter space into the performance space

function that maps points from the parameter space into a performance space,
Fig. 2.24.

Since the parameter space is constrained into a feasibility space, conditioned by
the several limitations that parameters values can assume, such as a minimum or
maximum transistor gate length, the performance space will also be conditioned into
a feasibility space.

So far, no circuit specifications were considered to define the feasibility space, but
designing a circuit involves respecting and achieving several target values for some
key circuit performances. The definition for these target performance specifications
are represented in the circuit sizing task as a problem constraint that usually reduces
the original performance feasibility space.

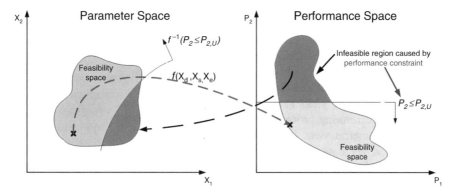

Fig. 2.25 A performance specification constraint reflects in the feasibility parameter space

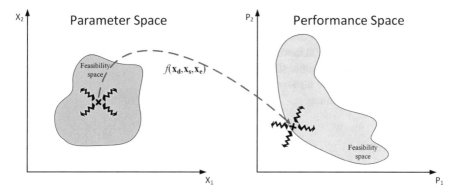

Fig. 2.26 Variability in parameter space causes variations in the performance values, resulting in parametric yield losses

Due to the relation between performance space and parameter space, constraints that affect performance space have an effect in the parameter space, because points in the parameters that mapped the new infeasible performance region are no longer accepted as feasible, Fig. 2.25.

The relation between parameter and performance space is typically nonlinear; thus, parameter solutions that safely exists near the center of the parameter feasibility region may correspond to performance space points dangerously close to the border of the performance feasibility region. Considering variability in the sizing process adds uncertainty to parameter's values, by adding perturbations into their nominal values. These perturbations are reflected in the performance space, causing variations in the performance values. In result of those variations some performance measures, once feasible, may cross into the infeasible region resulting in parametric yield losses, Fig. 2.26.

2.3.3 Parametric Yield

Two types of parameters variability are usually considered for improving circuit robustness and parametric yield. The first type of variability is related to manufacturing processes variations and the second relates to operating condition variations. Although parametric yield losses result from parameter variations caused by a nonideal manufacturing process, parametric yield is different from manufacturing yield. In parametric yield the circuits that do not comply with all the desired specifications are considered failures, whereas in manufacturing yield only the circuits that do not work at all are considered loses, defined as catastrophic manufacturing loses.

Manufacturing variability, caused by nonideal fabrication processes and nonidealities in materials, can be further classified into two categories: global variations, also known as inter-die variations, and local variations or intra-die variations, Fig. 2.27. Global variations are variations that can be defined between dies, wafer to wafer or, even, among wafer lots. Global parameter variations affect every device on the same die in identical manner, resulting in similar performance changes for each circuit instance. Since these parameters are independent, typically, they are represented as simple statistical distributions that model fluctuations around the nominal value of the parameter, which corresponds to the mean of the distribution with a given variance.

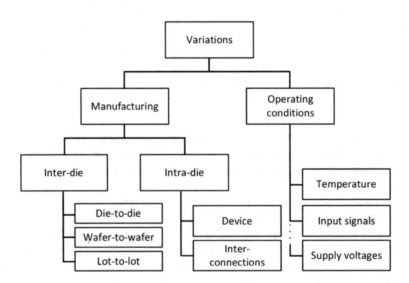

Fig. 2.27 Variations that affect parametric yield

Thus, global variations for a specific process parameter can be represented as:

$$P = P_{\text{nominal}} + \Delta P_{\text{global}} \tag{2.50}$$

where P_{nominal} is the nominal value of the process parameter and ΔP_{global} represents the global or inter-die variations as a random variable with zero mean.

The impact on circuit performances by a single process parameter can be estimated by process corner models and analyzed using deterministic worst-case methods. However, for a large number of process parameters variations and considering correlations among those parameters, the number of corner analysis grows exponentially, which makes the adoption of this approach limited. Corners are points in the variability space, defined by a combination of process parameters, temperature and voltages, that recreates some extreme working conditions or nonidealities in the fabrication process and materials.

Intra-die or local variations refer to parameter variations at different locations within a single die. Intra-die affects differently the devices planted on the same die, e.g., different gate oxide thickness in transistors. Thus, representing intra-die parameter variations requires separate random variables for each device in the circuit. In addition, spatial correlation among those random variables may exist. Devices located close to each other are expected to be affected in a similar way by some nonideal manufacturing process than devices located far away. Including local variations into (2.50), the total parameter variation expression is given by (2.51).

$$P_{\text{total}} = P_{\text{nominal}} + \Delta P_{\text{global}} + \Delta P_{\text{local}}(x, y) \tag{2.51}$$

where ΔP_{local} represents random local variations dependent on the spatial position (x, y) on the die.

Environmental operating conditions, such as temperature and supply voltages, under typical design phase of circuit sizing process are assumed to have fixed and invariant values. Since this assumption is unrealistic, operating conditions variations must be considered to forecast their effects on circuit specifications and, also, estimate circuit parametric yield. Operating conditions variations are modelled by intervals of parameters values, like a range of temperature operating condition [−5 °C, 100 °C], for which the circuit must guarantee that all the required specifications are still met.

Parametric yield requirements for today's analog ICs are quite demanding, values above 3σ are typical in this industry. A specification for parametric yield above 3σ means that more than 99.73% of the circuits must comply with required circuit specifications. Circuit specifications, typically, are presented in the form of one-sided constraints, i.e., DC gain ≥ 50 dB, power consumption ≤ 3 mW. Another type of specification constraint is referred as double-sided, where the specification is fulfilled for values in a bounded interval. Those two types of constraints define

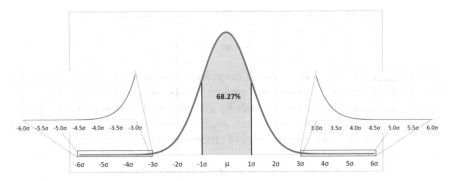

Fig. 2.28 Gaussian pdf with $\sigma = 1$

different yield values for the same sigma (σ) value, since the probability values are different. Also, sigma in this context refers to the standard deviation of the random specification's values obtained when the circuit is subject to random variability effects. Considering the large number of independent random variability effects in a circuit and using the central limit theorem it is possible to model the joint probability function of all variability effects as a Gaussian distribution [98].

Using a Gaussian distribution is easier to understand the relation among yield and sigma. In Fig. 2.28 a Gaussian probability density function (pdf), with zero mean ($\mu = 0$) and standard deviation equal to one ($\sigma = 1$), is depicted. As it is widely known Gaussian distributions assume the zero value only at $-\infty$ and $+\infty$, since the graphical representation of Fig. 2.28 wrongly presents the idea that for values close to $\pm 3\sigma$ the distribution is already zero, the regions from $\pm 3\sigma$ to $\pm 6\sigma$ are zoomed in. The region between $\mu - 1\sigma$ and $\mu + 1\sigma$, in light gray, corresponds to an area where the probability of finding a solution corresponds to P ($\mu - 1\sigma \leq X \leq \mu + 1\sigma$) = 68.269%. The relation among sigma and the probability of founding a solution in the area defined by the specification in terms of sigma can be used to define the yield value for a circuit. By performing simulations considering variability in a circuit solution, the mean and standard deviation are estimated. Then, it is possible to define the specification constraint in terms of the obtained sigma and using statistical tables that relate the sigma values with probability of the distribution, Table 2.4, and indirectly estimate the yield value.

The difference among one-sided constraint and double-sided is presented in Fig. 2.29. The one-sided constraint refers to a specification requirement for a value lower (greater) than some upper (lower) limit boundary value, like power consumption ≤ 3 mW or DC gain ≥ 50 dB, and where the limit boundary value corresponds to a multiple of sigma. Due to the symmetry characteristics of the Gaussian distribution it is possible to convert a specification requirement for values greater than some limit boundary value, e.g., DC gain ≥ 50 dB, and negative multiples of sigma limit values, into a similar situation as the one depicted in Fig. 2.29a by inverting the condition and signal of the value of the specification.

Table 2.4 Conversions among sigma values for one-sided and double-sided constraint based on the yield

One-sided constraint (σ)	Double-sided constraint (σ)	Estimated yield (%)
$-\infty$	0	0.0
-3	0.001691847	0.13498980
-1	0.200173687	15.86552539
0	0.674489753	50.0
0.475232849	1	68.26894921
1	1.409608713	84.13447461
2.782174975	3	99.73002039
3	3.205154926	99.86501020
5.88664517	6	99.99999980
6	6.113465073	99.99999990
$+\infty$	$+\infty$	100.0

The double-sided constraint represents a specification that are fulfilled for values between an interval, and the yield corresponds to the area under the curve limited by the values of sigma, Fig. 2.29b. For double-sided constraints where the sigma values are not symmetrical, it is possible to calculate the probability by simply subtracting the probability of the lower sigma to the probability of the higher sigma value, i.e., Yield $= P(\mu - 2\sigma \leq X \leq \mu + 3\sigma) = P(X \leq \mu + 3\sigma) - P(X \leq \mu - 2\sigma) = 99.865 - 2.275\% = 97.59\%$, Fig. 2.30.

2.4 Conclusion

Despite the evolution in automatic design of analog circuits, the fact is that analog IC design still represents an enormous effort in the overall chip design. To reduce the effort and improve productivity, designers add A/D converts as soon as possible in the path of the analog signals and perform most functions in digital circuitry. The reality is that this option is not always possible, which creates new challenges to EDA researchers and open new opportunities of business to the industry since IC technology keeps scaling down.

The different approaches for automatic analog IC sizing presented in this chapter will help frame the analog IC sizing tool adopted in this work. Furthermore, the discussed relation between circuit parameters and circuit performance spaces is required to understanding the yield estimation techniques described in the next chapter, as well as the new yield estimation methodology developed in this work.

In Table 2.5, a summary of the strengths and weaknesses of the previous discussed automatic analog IC sizing approaches is presented.

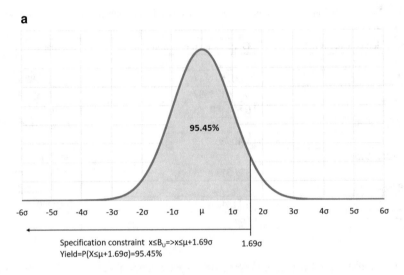

Specification constraint x≤B$_U$=>x≤μ+1.69σ
Yield=P(X≤μ+1.69σ)=95.45%

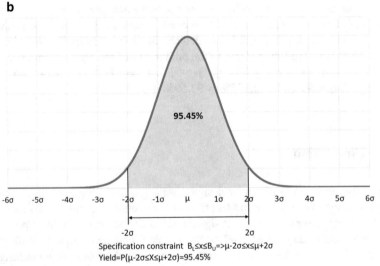

Specification constraint B$_L$≤x≤B$_U$=>μ-2σ≤x≤μ+2σ
Yield=P(μ-2σ≤X≤μ+2σ)=95.45%

Fig. 2.29 Estimating yield based on specification constraint in terms of σ. (**a**) One-sided specification constraint. (**b**) Double-sided constraint example

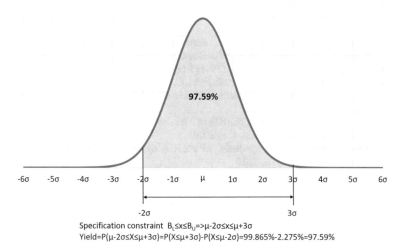

Specification constraint $B_L \leq x \leq B_U \Rightarrow \mu-2\sigma \leq x \leq \mu+3\sigma$
Yield$=P(\mu-2\sigma \leq X \leq \mu+3\sigma)=P(X \leq \mu+3\sigma)-P(X \leq \mu-2\sigma)=99.865\%-2.275\%=97.59\%$

Fig. 2.30 Non-symmetrical double-sided constraint

Table 2.5 Approaches applied in the analog sizing problem

		Optimization based		
	Knowledge based	Equations	Simulator	Models
Works	IDAC [19] OASYS [20] TAGUS [21–23] PAD [24, 25]	OPASYN [60] STAIC [61] del Mar Hershenson [63, 65, 66] Singh [67] Chen [62] ASTRX/OBLX [68] Meng [70] BA-ACO [71]	AIDA-C [75] Santos-Tavares [76] MAELSTROM [77] Anaconda [45] ASTRX/OBLX [68] Gupta [78] Dehbashian [80]	Yelten [81] Liu [82] You [89] Yengui [91] Bernardinis [94] Barros [95] Alpaydin [96]
Strengths	Fast execution time. Incorporates expert knowledge.	Fast execution time. Use of expert knowledge. Automatic symbolic analysis.	Models are easy to develop. Accurate and flexible.	Can be accurate Can be flexible
Weaknesses	Expert knowledge is difficult to capture. Not optimal.	Difficult derivation of some equations. Simplifications lead to lack of accuracy.	Long execution time. Limited to cell-level.	Accuracy is hard to predict.

References

1. F. Medeiro, R. Rodríguez-Macías, F. Fernández, R. Domínguez-Castro, J. Huertas, A. Rodríguez-Vázquez, Global design of analog cells using statistical optimization techniques. Analog Integr. Circ. Sig. Process **6**(3), 179–195 (1994)
2. E. Nowak, I. Aller, T. Ludwig, K. Kim, R. Joshi, C.-T. Chuang, K. Bernstein, R. Puri, Turning silicon on its edge [double gate CMOS/FinFET technology]. IEEE Circuits Devices Mag. **20**(1), 20–31 (2004)
3. B. Swahn, S. Hassoun, Gate sizing: finFETs vs 32nm bulk MOSFETs, in *2006 43rd ACM/IEEE Des. Automat. Conf.*, San Francisco, CA, 2006
4. R.A. Rutenbar, Analog layout synthesis: what's missing? in *Proc. 19th Int. Symp. Physical Des.*, San Francisco, CA, USA, 2010
5. B. Dobkin, J. Williams, *Analog Circuit Design*, Immersion in the Black Art of Analog Design, vol 2 (Newnes, Amsterdam, 2013)
6. R. Lourenço, N. Lourenço, N. Horta, *AIDA-CMK: Multi-Algorithm Optimization Kernel Applied to Analog IC Sizing* (Springer International Publishing, Cham, 2015)
7. G.G.E. Gielen, R.A. Rutenbar, Computer-aided design of analog and mixed-signal integrated circuits. Proc. IEEE **88**(12), 1825–1854 (2000)
8. A. Gerlach, J. Scheible, T. Rosahl, F. Eitrich, A generic topology selection method for analog circuits demonstrated on the OTA example, in *2015 11th Conf. Ph.D. Research in Microelectronics and Electronics (PRIME)*, Glasgow, 2015
9. P. Veselinovic, D. Leenaerts, W. van Bokhoven, F. Leyn, F. Proesmans, G. Gielen, W. Sansen, A flexible topology selection program as part of an analog synthesis system, in *Proc. 1995 European Conf. Design and Test*, Washington, DC, USA, 1995
10. A. Gerlach, T. Rosahl, F.-T. Eitrich, J. Scheible, A generic topology selection method for analog circuits with embedded circuit sizing demonstrated on the OTA example, in *2017 Des. Automat. Test Eur. Conf. Exhibition (DATE)*, Lausanne, Switzerland, 2017
11. H.E. Graeb, *Analog Design Centering and Sizing* (Springer, Dordrecht, 2007)
12. N. Lourenço, R. Martins, N. Horta, *Automatic Analog IC Sizing and Optimization Constrained with PVT Corners and Layout Effects* (Springer International Publishing, Cham, 2017)
13. R. Schwencker, F. Schenkel, M. Pronath, H. Graeb, Analog circuit sizing using adaptive worst-case parameter sets, in *2002 Des. Automat. Test Eur. Conf. Exhibition (DATE)*, Paris, France, 2002
14. R. Martins, N. Lourenço, A. Canelas, N. Horta, Stochastic-based placement template generator for analog IC layout-aware synthesis. Integration VLSI J. **58**, 485–495 (2017)
15. R. Martins, N. Lourenço, N. Horta, *Analog Integrated Circuit Design Automation – Placement, Routing and Parasitic Extraction Techniques* (Springer International Publishing, Cham, 2017)
16. B. Cardoso, R. Martins, N. Lourenço, N. Horta, AIDA-PEx: accurate parasitic extraction for layout-aware analog integrated circuit sizing, in *2015 11th Conf. Ph.D. Res. Microelectron. Electron. (PRIME)*, Glasgow, 2015
17. Tanner Calibre One Suite, Mentor Graphics, a Siemens Business. [Online]. Available: https://www.mentor.com/tannereda/calibre-one. Accessed 12 Nov 2018
18. M. Barros, J. Guilherme, N. Horta, *Analog Circuits and Systems Optimization Based on Evolutionary Computation Techniques* (Springer-Verlag, Berlin, 2010)
19. M. Degrauwe, O. Nys, E. Dijkstra, et al., IDAC: an interactive design tool for analog CMOS circuits. IEEE J. Solid-State Circuits **22**(6), 1106–1116 (1987)
20. R. Harjani, R. Rutenbar, L. Carley, OASYS: a framework for analog circuit synthesis. IEEE Trans. Comput. Aided Des. Integr. Circuits Syst. **8**(12), 1247–1266 (1989)
21. N. Horta, J.E. Franca, High-level data conversion synthesis by symbolic methods, in *IEEE Int. Symp. on Circuits and Systems. ISCAS 96*, Atlanta, GA, 1996
22. N. Horta, J.E. Franca, Algorithm-driven synthesis of data conversion architectures. IEEE Trans. Comput. Aided Des. Integr. Circuits Syst. **16**(10), 1116–1135 (1997)

23. N. Horta, Analogue and mixed-signal systems topologies exploration using symbolic methods. Analog Integr. Circuits Sig. Process. **31**(2), 161–176 (2002)
24. D. Stefanovic, M. Kayal, M. Pastre, V.B. Litovski, Procedural analog design (PAD) tool, in *Proc. 4th Int. Symp. on Quality Electronic Design*, 2003
25. D. Stefanovic, M. Kayal, *Structured Analog CMOS Design*, 1st edn. (Springer, Dordrecht, 2008)
26. F. El-Turky, E.E. Perry, BLADES: an artificial intelligence approach to analog circuit design. IEEE Trans. Comput. Aided Des. Integr. Circuits Syst. **8**(6), 680–692 (1989)
27. C. Toumazou, C.A. Makris, Analog IC design automation. I. Automated circuit generation: new concepts and methods. IEEE Trans. Comput. Aided Des. Integr. Circuits Syst. **14**(2), 218–238 (1995)
28. C.A. Makris, C. Toumazou, Analog IC design automation. II. Automated circuit correction by qualitative reasoning. IEEE Trans. Comput. Aided Des. Integr. Circuits Syst. **14**(2), 239–254 (1995)
29. C. Faragó, A. Lodin, R. Groza, An operational transcondutance amplifier sizing methodology with genetic algorithm-based optimization. Acta Technica Napocensis. Electronica-Telecomunicatii **55**(1), 15–20 (2014)
30. Cadence Design Systems, Inc., Virtuoso analog design environment family, 2014. [Online]. Available: https://www.cadence.com/content/dam/cadence-www/global/en_US/documents/tools/custom-ic-analog-rf-design/virtuoso-analog-design-fam-ds.pdf. Accessed 13 Sept 2018
31. MunEDA GmbH, muneda.com—Solutions optimization overview, 2018. [Online]. Available: https://muneda.com/solutions_optimization_overview.php. Accessed 13 Sept 2018
32. D. Payne, SemiWiki.com—Analog circuit optimization, 18 Apr 2012. [Online]. Available: https://www.semiwiki.com/forum/content/1194-analog-circuit-optimization.html. Accessed 13 Sept 2018
33. Synopsys, Inc., Synopsys completes acquisition of magma design automation, 22 Feb 2012. [Online]. Available: https://news.synopsys.com/index.php?s=20295&item=123356. Accessed 13 Sept 2018
34. H. Graeb, S. Zizala, J. Eckmueller, K. Antreich, The sizing rules method for analog integrated circuit design, in *IEEE/ACM Int. Conf. Comput. Aided Des. ICCAD 2001*, San Jose, CA, USA, 2001
35. T. Massier, H. Graeb, U. Schlichtmann, The sizing rules method for CMOS and bipolar analog integrated circuit synthesis. IEEE Trans. Comput. Aided Des. Integr. Circuits Syst. **27**(12), 2209–2222 (2008)
36. S. Kiranyaz, T. Ince, M. Gabbouj, Optimization techniques: an overview, in *Multidimensional Particle Swarm Optimization for Machine Learning and Pattern Recognition*, (Springer, Berlin, 2014), pp. 13–44
37. E. Tas, M. Memmedli, Near optimal step size and momentum in gradient descent for quadratic functions. Turk. J. Math. **41**(1), 110–121 (2017)
38. G.K. Wen, M. Mamat, I.B. Mohd, Y. Dasril, A novel of step size selection procedures for steepest descent method. Appl. Math. Sci. **6**(51), 2507–2518 (2012)
39. T.C. Hu, A.B. Kahng, *Linear and Integer Programming Made Easy* (Springer International Publishing, Cham, 2016)
40. W.L. Winston, *Operations Research: Applications and Algorithms*, 4th edn. (Brooks/Cole—Thomson Learning, Belmont, CA, 2004)
41. S. Boyd, S.-J. Kim, L. Vandenberghe, A. Hassibi, A tutorial on geometric programming. J. Optim. Eng. **8**(1), 67–127 (2007)
42. S. Kirkpatrick, C.D. Gelatt, M.P. Vecchi, Optimization by simulated annealing. Science **220**(4598), 671–680 (1983)
43. Y. Nourani, B. Andresen, A comparison of simulated annealing cooling strategies. J. Phys. A Math. Gen. **31**(41), 8373–8385 (1998)
44. V. Torczon, On the convergence of pattern search algorithms. SIAM J. Optim. **7**(1), 1–25 (1997)

45. R. Phelps, M. Krasnicki, R.A. Rutenbar, L.R. Carley, J.R. Hellums, Anaconda: simulation-based synthesis of analog circuits via stochastic pattern search. IEEE Trans. Comput. Aided Des. Integr. Circuits Syst. **19**(6), 703–717 (2000)
46. J. Kennedy, R. Eberhart, Particle swarm optimization, in *Proc. ICNN'95—Int. Conf. Neural Networks*, Perth, WA, Australia, 1995
47. M. Juneja, S.K. Nagar, Particle swarm optimization algorithm and its parameters: a review, in *2016 Int. Conf. Contr., Comput., Commun. Mater. (ICCCCM)*, Allahabad, 2016
48. M. Dorigo, M. Birattari, T. Stutzle, Ant colony optimization. IEEE Comput. Intell. Mag. **1**(4), 28–39 (2006)
49. S.-C. Chu, H.-C. Huang, J.F. Roddick, J.-S. Pan, Overview of algorithms for swarm intelligence, in *Comput. Collective Intell. Technol. Appl.: 3rd Int. Conf., ICCCI 2011*, Gdynia, Poland, 21–23 Sept 2011, *Proc., Part I*, ed. by P. Jędrzejowicz, N.T. Nguyen, K. Hoang (2011)
50. C. Blum, Ant colony optimization: introduction and recent trends. Phys. Life Rev. **2**(4), 353–373 (2005)
51. E. Rashedi, H. Nezamabadi-Pour, S. Saryazdi, GSA: a gravitational search algorithm. Inf. Sci. **179**(13), 2232–2248 (2009)
52. J.H. Holland, *Adaptation in Natural and Artificial Systems: An Introductory Analysis with Applications to Biology, Control and Artificial Intelligence* (MIT Press, Cambridge, MA, 1992)
53. K. Jebari, M. Madiafi, Selection methods for genetic algorithms. Int. J. Emerg. Sci. **3**(4), 333–344 (2013)
54. K. Deb, A. Pratap, S. Agarwal, T. Meyarivan, A fast and elitist multiobjective genetic algorithm: NSGA-II. IEEE Trans. Evol. Comput. **6**(2), 182–197 (2002)
55. M. Reyes-sierra, C.A.C. Coello, Multi-objective particle swarm optimizers: a survey of the state-of-the-art. Int. J. Comput. Intell. Res. **2**(3), 287–308 (2006)
56. Q. Zhang, H. Li, MOEA/D: a multiobjective evolutionary algorithm based on decomposition. IEEE Trans. Evol. Comput. **11**(6), 712–731 (2007)
57. K. Deb, H. Jain, An evolutionary many-objective optimization algorithm using reference-point-based nondominated sorting approach, part I: solving problems with box constraints. IEEE Trans. Evol. Comput. **18**(4), 577–601 (2014)
58. S. Chand, M. Wagner, Evolutionary many-objective optimization: a quick-start guide. Surv. Oper. Res. Manag. Sci. **20**(2), 35–42 (2015)
59. D.K. Saxena, T. Ray, K. Deb, A. Tiwari, Constrained many-objective optimization: a way forward, in *2009 IEEE Congr. Evol. Computation*, Trondheim, 2009
60. H.Y. Koh, C.H. Sequin, P.R. Gray, OPASYN: a compiler for CMOS operational amplifiers. IEEE Trans. Comput. Aided Des. Integr. Circuits Syst. **9**(2), 113–125 (1990)
61. J.P. Harvey, M.I. Elmasry, B. Leung, STAIC: an interactive framework for synthesizing CMOS and BiCMOS analog circuits. IEEE Trans. Comput. Aided Des. Integr. Circuits Syst. **11**(11), 1402–1417 (1992)
62. Y.L. Chen, W.R. Wu, C.N.J. Liu, J.C.M. Li, Simultaneous optimization of analog circuits with reliability and variability for applications on flexible electronics. IEEE Trans. Comput. Aided Des. Integr. Circuits Syst. **33**(1), 24–35 (2014)
63. M. del Mar Hershenson, S.P. Boyd, T.H. Lee, Optimal design of a CMOS op-amp via geometric programming. IEEE Trans. Comput. Aided Des. Integr. Circuits Syst. **20**(1), 1–21 (2001)
64. P. Mandal, V. Visvanathan, CMOS op-amp sizing using a geometric programming formulation. IEEE Trans. Comput. Aided Des. Integr. Circuits Syst. **20**(1), 22–38 (2001)
65. M. del Mar Hershenson, Design of pipeline analog-to-digital converters via geometric programming, in *IEEE/ACM Int. Conf. Comput. Aided Des.*, 2002
66. M. del Mar Hershenson, CMOS analog circuit design via geometric programming, in *Proc. Amer. Control Conf.*, Boston, MA, 2004
67. A.K. Singh, K. Ragab, M. Lok, C. Caramanis, M. Orshansky, Predictable equation-based analog optimization based on explicit capture of modeling error statistics. IEEE Trans. Comput. Aided Des. Integr. Circuits Syst. **31**(10), 1485–1498 (2012)

68. E.S. Ochotta, R.A. Rutenbar, L.R. Carley, Synthesis of high-performance analog circuits in ASTRX/OBLX. IEEE Trans. Comput. Aided Des. Integr. Circuits Syst. **15**(3), 273–294 (1996)
69. L.T. Pileggi, R.A. Rohrer, Asymptotic waveform evaluation for timing analysis. IEEE Trans. Comput. Aided Des. Integr. Circuits Syst. **9**, 352–366 (1990)
70. K.H. Meng, P.C. Pan, H.M. Chen, Integrated hierarchical synthesis of analog/RF circuits with accurate performance mapping, in *12th Int. Symp. Quality Electron. Des.*, Santa Clara, CA, 2011
71. B. Benhala, An improved ACO algorithm for the analog circuits design optimization. Int. J. Circuits Syst. Sig. Process. **10**, 126–133 (2016)
72. M. Fakhfakh, Y. Cooren, A. Sallem, M. Loulou, P. Siarry, Analog circuit design optimization through the particle swarm optimization technique. Analog Integr. Circuits Sig. Process. **63**(1), 71–82 (2010)
73. S. Kamisetty, J. Garg, J.N. Tripathi, J. Mukherjee, Optimization of analog RF circuit parameters using randomness in particle swarm optimization, in *2011 World Congr. Inform. Commun. Technol.*, Mumbai, 2011
74. A. El Dor, M. Fakhfakh, P. Siarry, Multiobjective differential evolution algorithm using crowding distance for the optimal design of analog circuits. J. Electr. Syst. **12**(3), 612–622 (2016)
75. N. Lourenço, R. Martins, A. Canelas, R. Póvoa, N. Horta, AIDA: layout-aware analog circuit-level sizing with in-loop layout generation. Integration VLSI J. **55**, 316–329 (2016)
76. R. Santos-Tavares, N. Paulino, J. Higino, J. Goes, J.P. Oliveira, Optimization of multi-stage amplifiers in deep-submicron CMOS using a distributed/parallel genetic algorithm, in *IEEE Int. Symp. Circuits Syst.*, Seattle, WA, 2008
77. M. Krasnicki, R. Phelps, R.A. Rutenbar, L.R. Carley, MAELSTROM: efficient simulation-based synthesis for custom analog cells, in *Proc. 1999 Des. Autom. Conf.*, New Orleans, LA, 1999
78. H. Gupta, B. Ghosh, Analog circuits design using ant colony optimization. Int. J. Electron. Comput. Commun. Technol. **2**(3), 9–21 (2012)
79. B. Benhala, O. Bouattane, GA and ACO techniques for the analog circuits design optimization. J. Theor. Appl. Inf. Technol. **64**(2), 413–419 (2014)
80. M. Dehbashian, M. Maymandi-Nejad, A new hybrid algorithm for analog ICs optimization based on the shrinking circles technique. Integration VLSI J. **56**, 148–166 (2017)
81. M.B. Yelten, T. Zhu, S. Koziel, P.D. Franzon, M.B. Steer, Demystifying surrogate modeling for circuits and systems. IEEE Circuits Syst. Mag. **12**(1), 45–63 (2012)
82. B. Liu, G. Gielen, F.V. Fernández, *Automated Design of Analog and High-Frequency Circuits* (Springer, Berlin, 2014)
83. R. Bellman, *Adaptive Control Processes: A Guided Tour* (Princeton University Press, Princeton, NJ, 1961)
84. H. Abdi, L.J. Williams, Principal component analysis. WIREs Comput. Stat. **2**(4), 433–459 (2010)
85. A. Giunta, S. Wojtkiewicz, M. Eldred, Overview of modern design of experiments methods for computational simulations, in *41st Aerospace Sci. Meeting Exhibit*, Reno, Nevada, 2003
86. H. Niederreiter, Random number generation and quasi-Monte Carlo methods. Soc. Ind. Appl. Math. (1992)
87. X. Wang, I.H. Sloan, Low discrepancy sequences in high dimensions: how well are their projections distributed? J. Comput. Appl. Math. **213**(2), 366–386 (2008)
88. C. Schlier, On scrambled Halton sequences. Appl. Numer. Math. **58**(10), 1467–1478 (2008)
89. H. You, M. Yang, D. Wang, X. Jia, Kriging Model combined with latin hypercube sampling for surrogate modeling of analog integrated circuit performance, in *2009 10th Int. Symp. Quality Electron. Des.*, San Jose, CA, 2009
90. G. Matheron, Principles of geostatistics. Econ. Geol. **58**(8), 1246–1266 (1963)

91. F. Yengui, L. Labrak, P. Russo, F. Frantz, N. Abouchi, Optimization based on surrogate modeling for analog integrated circuits, in *2012 19th IEEE Int. Conf. Electron. Circuits Syst. (ICECS 2012)*, Seville, 2012
92. M. Buhmann, *Radial Basis Functions: Theory and Implementations* (Cambridge University Press, Cambridge, 2003)
93. W. Hendrickx, T. Dhaene, Sequential design and rational metamodelling, in *Proc. Winter Simulation Conf.*, Orlando, FL, 2005
94. F.D. Bernardinis, M.I. Jordan, A.S. Vincentelli, Support vector machines for analog circuit performance representation, in *Proc. Des. Autom. Conf.*, Anaheim, CA, 2003
95. M. Barros, J. Guilherme, N. Horta, GA-SVM feasibility model and optimization, in *ACM Great Lakes Symp. VLSI*, Stresa-Lago Maggiore, 2007
96. G. Alpaydin, S. Balkir, G. Dundar, An evolutionary approach to automatic synthesis of high-performance analog integrated circuits. IEEE Trans. Evol. Comput. **7**(3), 240–252 (2003)
97. K.H. Lee, *First Course on Fuzzy Theory and Applications* (Springer Science & Business Media, Berlin, 2006)
98. T. McConaghy, K. Breen, J. Dyck, A. Gupta, *Variation-Aware Design of Custom Integrated Circuits: A Hands-on Field Guide* (Springer, New York, 2013)

Chapter 3
Yield Estimation Techniques Related Work

3.1 Yield Estimation Techniques

Variations introduced by nonideal IC manufacturing processes may affect the desired behavior of fabricated circuits. The prediction and maximization of the number of circuits that perform correctly, i.e., the circuit parametric yield, is an important factor to improve IC production profitability. This fact led to the appearance of several approaches that try to predict the circuit yield in early stages of the design flow. This section presents a brief introduction to yield estimation and to several works developed by the academia for yield estimation.

3.1.1 Parametric Yield Definition

In order to define the parametric yield estimation, consider that the circuit implements a function that maps the variable or parameter space X into the circuit performance space P, which can be represented as $P = f(\mathbf{x})$, Fig. 3.1. The performance space is described by the different performance parameters of interest, while the variable space is described by vector $\mathbf{x} = [x_{d, 1} \ldots x_{d, n}, x_{s, 1} \ldots x_{s, m}, x_{e, 1} \ldots x_{e, k}]^T = [x_1 \ldots x_{n + m + k}]^T \in \mathbb{R}^{n + m + k}$, joining the three types of parameters previously described in Sect. 2.3, i.e., design parameters $\mathbf{x}_d = [x_{d, 1} \ldots x_{d, n}]^T \in \mathbb{R}^n$, statistical parameters $\mathbf{x}_s = [x_{s, 1} \ldots x_{s, m}]^T \in \mathbb{R}^m$, and environment parameters $\mathbf{x}_e = [x_{e, 1} \ldots x_{e, k}]^T \in \mathbb{R}^k$.

Based on the optimization problem performance constraints it is possible to define a feasible performance space A_p that has a corresponding feasibility space in the variable design space A_s, where all desired circuit performances are fulfilled. The main objective of parametric yield optimization can be defined as finding a set of design parameters $\mathbf{x}_d$, that maximizes the probability of the solution is in the feasibility space A_s, considering the variations introduced by environment and

© Springer Nature Switzerland AG 2020
A. M. L. Canelas et al., *Yield-Aware Analog IC Design and Optimization in Nanometer-scale Technologies*, https://doi.org/10.1007/978-3-030-41536-5_3

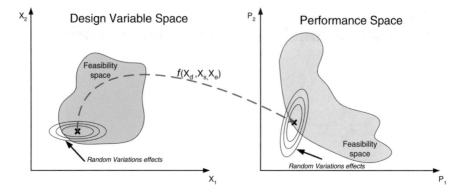

Fig. 3.1 Circuit function maps the variable design space into the performance space

process statistical parameters defined by the joint probability density function, pdf $(\mathbf{x}_s)$. The parametric yield can be given by integrating the probability density function over the feasibility space (3.1):

$$Y = \int_{A_s} \mathrm{pdf}(\mathbf{x}_s)d\mathbf{x}_s \tag{3.1}$$

In order to compute the integral in (3.1), the yield can be expressed in terms of all the space, as in (3.2).

$$Y = \int_{-\infty}^{+\infty} \mathbf{1}_{A_s}(\mathbf{x}_d, \mathbf{x}_s, \mathbf{x}_e)\, \mathrm{pdf}(\mathbf{x}_s)d\mathbf{x}_s \tag{3.2}$$

where $\mathbf{1}_{A_s}(x)$ corresponds to the indicator function:

$$\mathbf{1}_{A_s}(z) = \begin{cases} 1 & \text{if } z \in A_s \\ 0 & \text{if } z \notin A_s \end{cases} \tag{3.3}$$

The yield can be expressed as the expected value of $\mathbf{1}_{A_s}(\mathbf{x}_d, \mathbf{x}_s, \mathbf{x}_e)$:

$$Y = E_{\mathrm{pdf}(\mathbf{x}_s)}\left[\mathbf{1}_{A_s}(\mathbf{x}_d, \mathbf{x}_s, \mathbf{x}_e)\right] \tag{3.4}$$

The expected value in (3.4) is, usually, impossible to calculate since pdf$(\mathbf{x}_s)$ is unknown. To compute the integral is possible to reach for MC analysis and produce a random sequence of samples, which allows estimating the yield with an unbiased estimator for the expected value (3.5):

$$\widehat{Y} = \frac{1}{N} \sum_{i=1}^{N} \mathbf{1}_{A_s}\left(\mathbf{x}_d, \mathbf{x}_s^{(i)}, \mathbf{x}_e\right) = \frac{n_{\text{ok}}}{N} \tag{3.5}$$

where N is the total number of samples of the tested circuit solution.

Since for a fixed set of design variables only statistical parameters cause variations in performances, the samples are randomly generated based on the pdf($\mathbf{x}_s$), where the $\mathbf{x}_s^{(i)}$ denotes the i random sample. Considering the indicator function it is possible to evaluate the number of samples, n_{ok}, that meet all circuit specifications. Then, the yield is the percentage of samples that comply with the circuit desired specifications.

3.1.2 Monte Carlo Analysis for Parametric Yield Estimation

Typical MC analysis, for simulation-based approaches, samples are produced and evaluated by the electrical simulator. These samples are perturbations around the nominal variables' values of a circuit solution according to the foundry statistical models. The number of samples and the estimated yield value affect the accuracy of the estimator, according to the variance expression (3.6) if the yield value is known, and (3.7) if the yield is estimated [1].

$$\sigma_{\widehat{Y}}^2 = \frac{Y(1-Y)}{N} \tag{3.6}$$

$$\widehat{\sigma}_{\widehat{Y}}^2 = \frac{\widehat{Y}\left(1-\widehat{Y}\right)}{N-1} \tag{3.7}$$

The reduction of the variance, for a more accurate yield estimation result, implies testing a larger number of random samples. Considering a yield of 85% and 50 samples, from (3.6), since the yield is known, results a standard deviation of 5%. Using a $3\sigma_Y$ interval for the 85% of yield follows that the yield is in the interval [70%, 100%] with a probability of 99.73%. By increasing the number of samples to a size of 500, the standard deviation reduces to 1.6%; thus, considering also $3\sigma_Y$, the yield is in an interval [80.2%, 89.8%]. The increase in the number of samples reduces the standard deviation of estimated yield which increases the confidence level in the estimated result, Table 3.1.

Following the assumption to deduce (3.6), that the samples are independently and identically distributed, and in number enough to approximate the yield estimation distribution to a normal distribution, it is possible to easily find the number of samples required to estimate the yield with a certain confidence interval. The confidence level, C_y, is the probability that the estimated yield value is on some given interval, $C_y = P(Y - k_Y\sigma_Y \leq Y \leq Y + k_Y\sigma_Y)$, where the interval can be expressed in multiples of the standard deviation distribution. Since a normal

Table 3.1 Standard deviation of the yield estimator and $3\sigma_Y$ intervals for a yield of 85% and for a different number of samples

N	10	50	100	500	1000	5000
σ_Y	11.3%	5.0%	3.6%	1.6%	1.1%	0.5%
$3\sigma_Y$ interval (%)	[51.1, 100]	[70, 100]	[74.2, 95.8]	[80.2, 89.8]	[81.7, 88.3]	[83.5, 86.5]

Table 3.2 Number of samples required according to the desired confidence level and confidence interval for 90% and 95% estimated yield values

Confidence level	68.27%		95.45%		99.73%		99.99%	
$k_Y\sigma_Y$	$1\sigma_y$		$2\sigma_y$		$3\sigma_y$		$6\sigma_y$	
Yield values	90%	95%	90%	95%	90%	95%	90%	95%
ΔY	Number of samples—N							
±10%	9	5	36	19	81	43	324	171
±5%	36	19	144	77	324	171	1296	684
±1%	900	475	3600	1900	8100	4275	32,400	17,100

distribution is being considered, it is possible to adopt the idea introduced at the end of Chap. 2, about the relation between probability value and the value of σ, to consider that for a confidence level $C_y = 90\%$ the estimated yield value is in an interval of $\pm 1.645\sigma_y$. Using the afore-mentioned relation, finding the k_Y value according to some desired confidence level is an easy task. Defining the confidence interval as $Y \pm \Delta Y$, where $\Delta Y = k_Y\sigma_Y$, allows eliminating the standard deviation from (3.6), and obtain an approximate number of samples required for the desired confidence level (3.8):

$$N = \left\lceil \frac{Y(1 - Y)k_Y^2}{\Delta Y^2} \right\rceil \tag{3.8}$$

where the function $\lceil x \rceil$ rounds up to the nearest integer number.

Using (3.8) allows computing the number of samples required for a desired estimated yield value confidence level and interval, Table 3.2. From Table 3.2 it is possible to deduce the relation between accuracy and the number of samples. The presented values show that for an increase in accuracy of 10 times, from $\Delta Y = \pm 10\%$ to $\Delta Y = \pm 1\%$ in the confidence interval, the number of samples has to grow approximately 100 times. Therefore, in MC the accuracy is proximately proportional to the square root of the number of samples, Accuracy $\propto \sqrt{N}$.

Table 3.2 results present the number of samples for two estimated yield values, 90% and 95%. The computed sample number shows that the estimated yield value of 90% always requires more samples than the 95% value for the same confidence level and interval. This fact can be explained by considering the first (3.9) and second derivative (3.10) of the variance of yield estimator (3.6), with respect to Y:

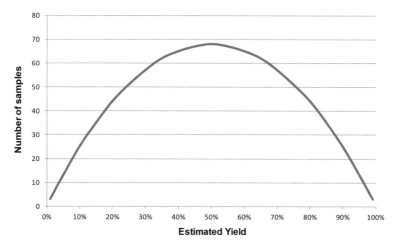

Fig. 3.2 Number of samples required to maintain the same estimation accuracy and confidence level for different yield values

$$\frac{\partial \sigma_{\hat{Y}}^2}{\partial Y} = \frac{1 - 2Y}{N} \tag{3.9}$$

$$\frac{\partial^2 \sigma_{\hat{Y}}^2}{\partial Y^2} = -\frac{2}{N} \tag{3.10}$$

The second derivative in (3.10) is always negative and the yield value that zeros the first derivative is $Y = 50\%$, so the variance is maximum for an estimated yield value of 50%, which is why the number of samples required for the 95% estimated yield value is smaller than the 90% yield with the same accuracy.

In Fig. 3.2, the fact that the estimated yield variance has a maximum at 50% is depicted. The graphic shows that to estimate the yield with the same accuracy a larger number of samples is required for that yield value, (50%). The values used to generate the graphic, where the yield span from 1% to 99%, bounded by $0\% \leq Y - \Delta Y$ and $Y + \Delta Y \leq 100\%$, were a confidence level of 90%, thus $k_Y = 1.645$ and $\Delta Y = 10\%$.

3.1.3 Yield Estimation Methodologies

The reduction of the variance, for a more accurate yield estimation result, implies testing a larger number of random samples, as was discussed earlier. This fact led to new approaches that try to minimize the time impact of testing a higher number of samples. Several of those approaches are detailed next.

3.1.3.1   Monte Carlo or Quasi Monte Carlo Sample Methods

In typical MC for yield estimation, the samples are randomly, or pseudo-randomly, chosen. Other approaches like QMC adopt low discrepancy sampling methods that try to evenly sample the entire parameter space. Different sampling techniques are also available in commercial electrical simulators, like in Mentor Eldo©, which offer random sampling, QMC, and LHS to perform MC analysis.

The question of which sampling technique adopts to statistical circuit analysis does not have a unique answer. Several works, described later in this section, adopt QMC, others LHS. Works that model circuit performance can implement typical MC pseudo-random sampling (MCS) since circuit performance evaluations are quite fast. In [2], Singhee et al. compares QMC with MCS and with LHS. Comparing QMC with MCS revealed that QMC has a better convergence rate $[O(1/n)]$ than typical MCS $[O(1/\sqrt{n})]$, which allows using a smaller number of samples for the same accuracy. Results showed a $2\times$ to $8\times$ speedup performance for roughly 1% accuracy level. The tests were performed using three circuits, at nanometer technology nodes, and showed that LHS is between QMC and MCS in terms of time performance and convergence rate, being the best approach the QMC.

Liu et al., in [3], presented the ordinal optimization-based random-scale differential evolution (ORDE) algorithm, which intends sizing and optimizing analog circuits considering variability. The adopted optimization kernel is based on the differential evolution-based (DE) algorithm, discussed earlier. Since DE is a population-based algorithm, to estimate the yield of each individual a large number of simulations considering variability must be performed on each solution. To reduce the time impact related to performing a large number of simulations at each iteration of the algorithm, the authors adopted a group of techniques. First infeasible solutions at typical simulations are discarded, because it does not make sense waste expensive variability-simulation time on individuals that are unable to comply with some performance constraint at ideal typical conditions. The second technique uses ordinal optimization (OO) [4], which is based on the idea that it is much easier to determine "order" than "value" to rank solutions, and instead of looking for "the best for sure" seek the "good enough with high probability." In the two-phase yield estimation process, the OO algorithm first orders the feasible nominal solutions at each iteration based on a rough yield value, estimated using a small number of MC simulations, and according to the defined ranking allocates an available budget of MC simulations to the different solutions, where a higher number of simulations is allocated to the best solutions or to solutions with high variance in the circuit performances of interest, in comparison to the best solutions, meaning that the accuracy of the yield estimation of those solutions is low, so a higher number of simulations must be performed to improve accuracy. The third technique to reduce the impact of a large number of MC simulations is adopting LHS. By using LHS, the number of simulations to estimate the yield, with a confidence interval of 1% and a 90% confidence level, was reduced to 20%, when compared to classic MC. The OO algorithm works as an optimization loop inside the DE-based optimization kernel to

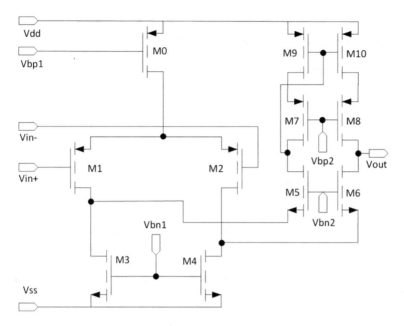

Fig. 3.3 Folded cascade operational transconductance amplifier tested by Guerra-Gomez et al. [5]

estimate the circuit yield measure, which is then used by the DE-based algorithm to size the circuit and optimize some important circuit performances considering the variability.

A similar MC simulations budget allocation process was adopted by Guerra-Gomez et al. in [5]. The optimal computing budget allocation (OCBA) strategy is used by Guerra-Gomez et al. to distribute the large portion of the simulation budget among critical potential solutions and limit the number of simulations for the noncritical solutions. The OCBA strategy was implemented in three different multi-objective optimization algorithms, the MOEA/D, the NSGA-II, and the MOPSO. Two genetic operators were tested on NSGA-II and the MOEA/D: the simulated binary crossover (SBX) [6] and differential evolution (DE). The tests were carried out using just one circuit, a folded cascade operational transconductance amplifier in a 90 nm technology node, Fig. 3.3, and the circuit performances were evaluated by HSPICE©.

Guerra-Gomez et al. defined the optimization problem constraints so that all transistors must operate in the saturation region, considering these constraints as strong specification constraints. Additional target specification conditions were considered as weak constraints. The problem design variables are the gate width and length of all CMOS transistors in the circuit, adding up to seven variables due to transistor matching. The optimization problem objectives are the maximization of the gain DC, the gain bandwidth, and slew-rate, and, also, the minimization of the input referred noise, input offset voltage, settling time, and power consumption.

The three evolutionary-based optimization algorithms tested by Guerra-Gomez et al., i.e., MOEA/D, NSGA-II, and MOPSO, adopt a population size of 210 individuals and run along 250 generations. Considering just the final yield values, the algorithm with better results was the MOPSO using the OCBA strategy reaching 90% of yield, the NSGA-II algorithms provide yield solutions below 73% and the MOEA/D was unable to provide solutions better than 54% of yield. The algorithm that showed a large reduction in the number of simulations, when compared to classical MC with 100 iterations per solution, was the MOPSO with 85% of reduction, while the MOEA/D with SBX reduced the number of simulations in 55% and with DE crossover in 59%. For the NSGA-II with SBX the reduction in simulations was 25% and with DE was near 0.5%. Most results are unacceptable in terms of yield; solutions with 36% of yield like the one obtained by MOEA/D with DE and OCBA will never be fabricated. The authors did not provide much information about the tests; for instance it is not clear if more than one run was performed to test the different approaches, and the weak constraints defined in the optimization problem were not revealed.

In [7], Afacan et al. presented an analog optimization tool also using the OO algorithm with simulation budget allocation to estimate the yield as an optional module. Moreover, a technique was implemented to discard infeasible solutions from the MC simulations, which the authors called Infeasible Solution Elimination (ISE). The ISE module aims to decrease the circuit synthesis time by performing variability analysis only to potential solutions that satisfy some predefined specification constraints.

The adopted single-objective hybrid optimization kernel uses evolutionary strategies to create new potential solutions followed by a SA technique to guide the search for optimal solutions. The main yield estimation module adopts QMC sampling to estimate the yield of potential solutions. The yield estimation module implements a self-adaptive mechanism to stop MC simulations. This stopping criterion is activated once the Kullback-Leibler divergence between histograms of two consecutive sets of samples is below some threshold value.

The Kullback-Leibler divergence can be defined as (3.11).

$$D_{KL}(P \parallel Q) = \sum_i P(i) \ln \left(\frac{P(i)}{Q(i)} \right) \tag{3.11}$$

where $P(i)$ and $Q(i)$ are discrete probability distribution.

As a practical example of the stopping criterion, consider performing 50 simulations. Then, based on the results compute a histogram to estimate distribution $P(i)$. Next perform 50 more simulation and again compute the histogram for distribution $Q(i)$ using all 100 simulations. After, apply (3.11) to calculate the divergence, Fig. 3.4.

As similarity between consecutive histograms increase, the Kullback-Leibler divergence is reduced, so the authors defined a value for which both histograms are considered identical and the simulations to estimate the yield are interrupted, thus

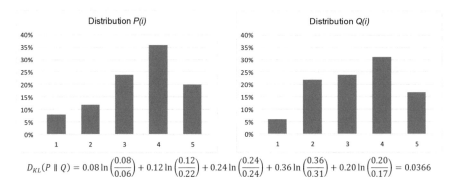

$$D_{KL}(P \parallel Q) = 0.08 \ln\left(\frac{0.08}{0.06}\right) + 0.12 \ln\left(\frac{0.12}{0.22}\right) + 0.24 \ln\left(\frac{0.24}{0.24}\right) + 0.36 \ln\left(\frac{0.36}{0.31}\right) + 0.20 \ln\left(\frac{0.20}{0.17}\right) = 0.0366$$

Fig. 3.4 Kullback-Leibler divergence example

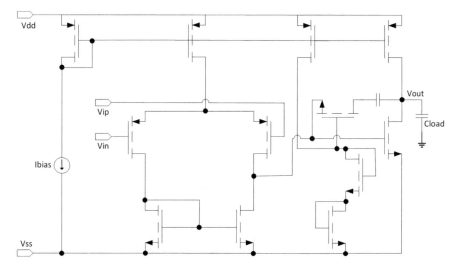

Fig. 3.5 Two-stage operational transcondutance amplifier tested by Afacan et al. in [7]

reducing the total number of simulations needed to estimate the yield. At the end of the optimization process the user has the ability to estimate the yield of each candidate solution using the OO algorithm with simulation budget allocation, which according to the authors provide a more accurate yield estimation. The developed tool is programmed in MATLAB© and the circuit performances were estimated using HSPICE©. Two circuits, Figs. 3.5 and 3.6, in a 130 nm technology node were used to evaluate the tool.

The final circuit solutions were compared with solutions from a standard sizing process with no variability concerns. The solutions obtained with the standard sizing process provide good circuit performances values with very low yield, while the circuits sized by the yield-aware tool achieved solutions up to 98% of yield, but with lower performances, as expected. The cost of considering variability is revealed in

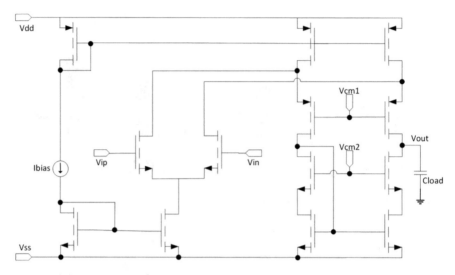

Fig. 3.6 Folded cascode amplifier tested by Afacan et al. in [7]

the optimization processes execution time. The standard process takes 15–20 min whereas the yield-aware tool takes between 60 and 70 min in an Intel© i7 chipset with 2.8 GHz CPU. The stopping criterion presents as a good technique to reduce the total number of simulations for yield estimation but dividing the samples in sets forces invoking multiple times the electrical simulator, which may result in additional execution time.

Afacan et al. in [8] presented a new approach for analog circuit sizing considering yield estimation. This new tool divides the yield estimation process in two phases. The first phase receives only feasible solution, filtered by an ISE process, and performs a low number of simulations using QMC to provide a coarse yield estimation. Then, solutions are examined in a new ISE process, which discards solutions with yield value below some predetermined yield threshold. The solutions that survive to the second ISE filter are submitted to a second process to estimate the yield. The second yield estimation phase adopts scrambled-QMC sampling, which is based on random permutations of samples sets. The adopted sampling approach in the second yield estimation phase, the scrambled-QMC [9], was selected to overcome a QMC limitation related to the fact that this sampling technique is based on deterministic sequences, so it has no natural variance which makes impossible to estimate its statistical error. Adopting scrambled-QMC is possible at the end of the second yield estimation phase to compute the standard deviation and the Y_{upper} upper and Y_{lower} lower bounds of the yield estimation. Using those bounds, the authors compute a coefficient, K_y, that later scales the solution fitness (3.12).

$$K_y = \frac{|Y_{\text{best}} - Y_{\text{lower}}| + \varepsilon}{Y_{\text{upper}} - Y_{\text{lower}}}, \quad \text{with } \varepsilon \cong 0.1 \tag{3.12}$$

The initial fitness or cost function is based on weighted sum of all objective performances and also includes a penalization component for transistor not operating on the desired region (3.13).

$$C = C_{\text{perf}} + C_{\text{pen}} \tag{3.13}$$

with,

$$C_{\text{perf}} = \sum_i^n w_i P_i^2 \tag{3.14}$$

$$P_i = \frac{U_i - f_i}{U_i - L_i}, \quad P_{i,\text{min}} = 0 \tag{3.15}$$

where n is the number of performance specifications, f_i is the value of performance specification i, and U_i and L_i are the upper and lower values of performance i, respectively. C_{pen} is calculated according to the operating point of all transistors, where triode and cutoff regions are penalized. A weight value w_i is specified for each performance specification, which is automatically increased for performances that are not satisfied.

After estimating the yield, the fitness value (3.13) is multiplied by the yield coefficient K_y, (3.12), which reduces the value of the fitness function according to the solution yield value. Thus, the new fitness function including the yield coefficient favors higher yield solutions, which guides the evolutionary optimization kernel to search for more robust solutions. At the end of the optimization process, a final MC run of 10,000 iterations is performed to accurately estimate the yield of the best solutions. The proposed yield-aware tool was tested using the same two test circuits of the previous work, Figs. 3.5 and 3.6, in a 130 nm technology node, and three optimization runs were executed. Like in his previous work, Afacan et al. compared this new approach with a standard sizing process without any variability concerns. The standard sizing process took 15–20 min to conclude and the yield-aware sizing process took 60–70 min in an Intel© i7 chipset with 2.8 GHz CPU. The final optimization results show circuits with yield values above 95%; this high yield values were only possible to achieve for circuits that do not comply with the power and area constraints.

Afacan et al. at their last version of the circuit synthesis tool with yield estimation, [10], presented a work that combines the previous presented techniques in [7, 8]. This version of the tool also implements a two-phase yield estimation technique preceded by two ISE stages and, also, adopts scrambled-QMC sampling. The major change with respect to [8] was the addition of the adaptive sample technique using the Kullback-Leibler divergence at the second yield estimation phase. The authors reported that the new tool version was tested using the same

Table 3.3 Experimental data for the selection of QMC sample size, from Pak et al. [11]

Sampling method	Number of samples	% Variation obj1/ obj2 solution 1	% Variation obj1/ obj2 solution 2	% Variation obj1/ obj2 solution 3
MCS	10,000	13.20/1.61	34.22/2.06	4.53/1.91
QMC	500	12.88/1.57	33.63/2.02	4.42/1.86
QMC	100	12.66/1.53	32.91/1.98	4.33/1.81
QMC	75	12.53/1.49	32.35/1.97	4.31/1.80
QMC	50	12.43/1.48	31.85/1.97	4.28/1.79
QMC	25	11.85/1.36	29.75/1.92	4.12/1.24

MCS pseudo-random sampling or MC sampling, *QMC* quasi-MC sampling

circuits as in [8], and the results revealed higher yield solutions, and, with better accuracy. But like in previous versions of the tool, many of the presented solutions do not respect area and power constraints, and with great surprise, the yield reported to those solutions is of 100%.

The work presented in [11] by Pak et al. implements and tests different techniques to consider the yield of analog ICs in a synthesis loop. MOEA/D was the optimization algorithm adopted to size the circuits and optimize circuit performances. The QMC sampling approach was adopted for the variability simulations. To select the number of QMC iterations to perform per simulation, a study using 10 different runs was performed and the results were compared with a MC run using 10,000 iterations, Table 3.3.

The comparison between the 10 QMC runs with the MC run revealed that 25–50 QMC iterations or samples are good enough for a coarse yield estimation, and for a more accurate yield estimation the number of iterations should increase to the range of hundreds. The authors presented Table 3.3 to support their conclusions, where three different solutions, selected from a two-dimensional Pareto front, were submitted to MC simulations using a different QMC number of samples. The "% *variation*" columns of Table 3.3 are the 10 runs average values of the variations of each Pareto objective for each selected solution. The closer the variation percentage value is to the reference MC sampling variation value, the most accurate the yield estimation is.

The presented yield-aware sizing tool tested three different techniques to incorporate the variability information into the sizing processes. The first technique, named cloud width (CW), corresponds to the sum of the distance between the nominal performance and worst-case performance and the distance of the nominal performance to the best-case performance, $CW = x_{wc} + x_{bc}$. The x_{wc} distance between the nominal performance and the worst-case performance corresponds to the maximum Euclidian distance computed from the nominal performance to every QMC simulated samples with a performance worse than the nominal value, which in a minimization problem is a value higher than the nominal value, (3.16).

$$x_{wc} = \max_{1 \leq j \leq N} \left\{ \sqrt{\sum_{i=1}^{m} \left(x_{i,j} - x_i^{\mathrm{nom}} \right)^2} \right\}, \quad \text{if} \left(x_{i,j} - x_i^{\mathrm{nom}} \right) > 0 \text{ for all } 1 \leq i$$
$$\leq m \text{ of each } j \tag{3.16}$$

$$x_{bc} = \max_{1 \leq j \leq N} \left\{ \sqrt{\sum_{i=1}^{m} \left(x_{i,j} - x_i^{\mathrm{nom}} \right)^2} \right\}, \quad \text{if} \left(x_{i,j} - x_i^{\mathrm{nom}} \right) < 0 \text{ for all } 1 \leq i$$
$$\leq m \text{ of each } j \tag{3.17}$$

where N is the number of samples; m is the number of performance specifications; $x_{i,j}$ performance i value of sample m and x_i^{nom} nominal performance i value.

The x_{bc} best-case to nominal performance is the maximum Euclidian distance between the nominal performances to every QMC simulated samples with a performance better than the nominal value, (3.17). In the optimization problem a constraint that uses the CW value was defined, where every solution with a CW value above the threshold value is considered infeasible. The second technique tested was the worst-case Pareto front (WCPF). The WCPF is based on the distance between the worst-case performance and the nominal performance value, defined like in the CW technique, $\mathrm{WCPF} = x_{wc}$. Using the worst-case distance, it is possible to identify the worst-case solutions and consider these solutions to the Pareto front. As a result, the Pareto front is defined by the worst-case solutions which lead the optimizer to try improving on these pessimistic performances. The last and third technique tested was the individual-based yield (IBY), (3.18). The IBY technique penalizes solutions whose yield value is below some predefined value. During tests, all solutions with yield below 80% had penalized their fitness.

$$\mathrm{IBY} = \frac{\sum_{j=1}^{N} Y_j}{N} \tag{3.18}$$

with,

$$Y_j = \begin{cases} 1, & \text{if } x_{i,j} < (1 - d_i) \cdot x_i^{\mathrm{nom}} \quad \text{for all } 1 \leq i \leq m \\ 0, & \text{otherwise} \end{cases} \tag{3.19}$$

where N is the number of samples; $x_{i,j}$ performance i value of sample m, x_i^{nom} nominal performance i value and d_i is a user-input relaxation parameter to accept performance i constraint.

The yield-aware sizing process, which took 5 h to complete for two objectives and 9 h for three objectives, was compared with a nominal sizing process, which required 20 min to conclude in the two-objective problem and 75 min for the three-objective case on a 1.9 GHz i3-3227 processor. For both problems, with two objectives and three objectives, the yield-aware optimizer started from a Pareto front resulted from a nominal optimization process. The tests showed that for the two-objective optimization problem all three techniques, i.e., CW, WCPF, and IBY, were able to provide similar results, whereas for the three-objective problem the IBY technique revealed worst results than the other two techniques.

3.1.3.2 Monte Carlo Importance Sampling Methods

The works presented so far reduce the time impact of MC simulations by means of LHS or QMC sampling, which permits using a smaller number of samples to estimate the yield than with typical MC. Another approach to reduce the number of samples is important sampling (IS). The IS technique is widely adopted in rare circuit failure events where typical MC is infeasible due to the very large number of MC simulations required; for instance, hundreds of millions of samples are needed to estimate a failure rate of 10^{-6} [12, 13]. IS tries to improve efficiency by shifting the sampling region towards critical or failure performance specification areas. Typically, IS is composed of two phases. At the first phase IS explores the sampling region to identify the failure region; this process is achieved by uniform sampling with a low number of samples or another linear fast approach. After finding the failure region, the second phase centers the new sampling distributions at the failure region and performs a large number of samples to estimate the yield. The final yield value is computed using the new samples, and each sample $\mathbf{s}_i$ is weighted, $w_i(\mathbf{s}_i)$, based on the probability density of the original distribution, $p(\mathbf{x})$, and the new sampling distribution, $g(\mathbf{x})$.

$$\mathbf{w} = [w_1, \ldots, w_N]; \quad \text{where } w_i(\mathbf{s}_i) = \frac{p(\mathbf{s}_i)}{g(\mathbf{s}_i)} \tag{3.20}$$

The weights are scaled to the range [0, 1] by (3.21).

$$v_i(\mathbf{s}) = \frac{w_i(\mathbf{s}_i)}{\sum_j w_j(\mathbf{s}_i)} \tag{3.21}$$

Using the weights, the yield is estimated by (3.22):

$$\widehat{Y} = \frac{1}{N} \sum_{i=1}^{N} v_i(\mathbf{s}_i) \mathbf{1}_P(\mathbf{s}_i) \tag{3.22}$$

where N is the number of samples, function $\mathbf{1}_P(\mathbf{s}_i)$ is 1 (one) if sample $\mathbf{s}_i$ is feasible, i.e., $\mathbf{s}_i$ complies with all performance requirements, and 0 (zero) otherwise.

The IS yield estimator in (3.22) can be deduced using the same reasoning from (3.2) to (3.5) considering:

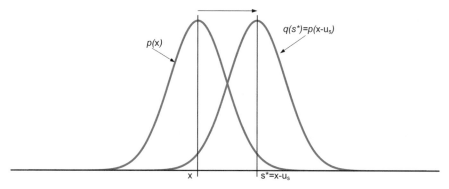

Fig. 3.7 Importance sampling shifts the sampling distribution to the point of failure

$$Y = \int_{-\infty}^{+\infty} 1_{A_s}(\mathbf{x}_d, \mathbf{x}_s, \mathbf{x}_e) \, \text{pdf}(\mathbf{x}_s) d\mathbf{x}_s$$

$$= \int_{-\infty}^{+\infty} 1_{A_s}(\mathbf{x}_d, \mathbf{x}_s, \mathbf{x}_e) \frac{\text{pdf}(\mathbf{x}_s)}{\text{pdf}^*(\mathbf{x}_s)} \text{pdf}^*(\mathbf{x}_s) d\mathbf{x}_s \qquad (3.23)$$

where the probability density functions are replaced according to $p(\cdot) = \text{pdf}(\cdot)$, $g(\cdot) = \text{pdf}^*(\cdot)$ and the weight term is $w_i(\cdot) = \text{pdf}(\cdot)/\text{pdf}^*(\cdot)$.

A critical step in IS is finding the point of interest, which is the point to center the new sampling distribution and the point of failure.

One of the most widely adopted approaches is shifting the mean of the original distribution to a point $\mathbf{s}^*$, where $\mathbf{s}^*$ is defined as $\mathbf{s}^* = \arg \min_{\mathbf{u}_s} \|\mathbf{x} - \mathbf{u}_s\|$ subject to $\mathbf{x} - \mathbf{u}_s \in \boldsymbol{\Phi}$, where $\boldsymbol{\Phi}$ is the failure region defined by all performance constraints, Fig. 3.7.

In [14], Kanj et al. presented an approach to find the point of interest using the mean-shift idea. In the first step of their approach, Kanj et al. used uniform sampling to explore the parameter space around point $\mathbf{x}$; the sampling process continues until more than 30 failure points are identified. Next, the center of gravity of the failure points is computed and this point becomes the point of interest $\mathbf{s}^*$. The adopted circuit to test the approach is a simple SRAM bit cell with six transistors, Fig. 3.8.

The SRAM bit cell circuit is used in most of the works dedicated to rare failure events due to the extremely low failure rate when considering the circuit by itself, but when a very large number of SRAM bit cells replicas are connected to build a memory chip the join failure rate starts to become important. The authors considered six independent Gaussian process variables, and based on the results concluded that performing 2000–3000 IS iterations/simulations for yield estimation is equivalent to performing 100k MC iterations in terms of accuracy. Later, in [15] McConaghy et al. tested this approach using 185 variables and concluded that Kanj et al. approach can handle from 6 to 12 variables but for a larger number of variables it is not good

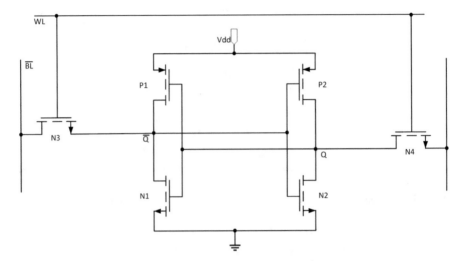

Fig. 3.8 6T SRAM bit cell schematics

enough to solve the rare failure event problem. The previously described IS approach considers only one failure region, which is the most likely failure region.

Yao et al. in [16] presented an approach using importance boundary sampling which is used to estimate the yield considering multiple failure regions. Importance boundary sampling instead of locating the most likely point of failure tries to identify all the most likely failure regions during the first stage. Next, at the second stage, the boundaries of each most likely failure region are identified and fully characterized. The boundaries are characterized by surrogate models. Finally, the IS is performed for each failure region, taking advantage of the surrogate models from stage 2, to estimate the failure rate without performing electrical expensive simulations. The approach was tested with the SRAM bit cell and the test showed a speedup of $5\times$–$20\times$ when compared to other techniques. Like the previous approach, this method suffers from the problem that it is only applicable to low-dimensional problems, where up to 30 process variables are considered.

Trying to improve the IS approach in order to be able to deal with high-dimensional problems and multiple failure regions, Wang et al. in [17] presented a new approach. The proposed approach is divided into two optimization loops (Fig. 3.9). The first global optimization loop finds multiple failure points in different regions and reduces its dimensionality by clustering them together. Next, the second optimization loop performs a local search using sequential quadratic programming (SQP), where the clusters centroids are used as starting point to find the optimal point of interest at each failure region.

To further improve computational efficiency, a surrogate model was developed to estimate circuit performances. The approach was tested using the typical SRAM bit cell using up to 384 dimensions. The results when compared with the importance boundary sampling (IBS), previously described, revealed that IBS was unable to identify failure regions even with 50,000 simulations, while for the same number of

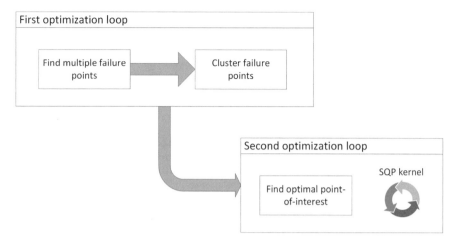

Fig. 3.9 Two-step optimization process adopted by Wang et al. [17]

simulations the presented approach was able to achieve what the authors called a "golden result," which is a result close the result obtained for typical MC yield estimation performing 1e8 simulations.

Based on the IS technique, Sun et al. in [18] presented the scaled-sigma sampling (SSS). In this technique the new sampling distribution, $g(\mathbf{x})$, is just a version of the original distribution $p(\mathbf{x})$ whose standard deviation was scaled by a factor s, with $s > 1$. As a result, the new sampling distribution has the same mean value of the original distribution, but due to the new sigma value, samples are generated farther away from the mean than in the original sampling distribution, making more probable to samples reach the failure region (Fig. 3.10).

The SSS technique was tested using the SRAM bit cell and compared with other IS techniques. The results show that this technique presents better results for problems with high-dimensional variation spaces, while for low-dimensional spaces, e.g., 6–20 random variable, the other IS techniques presented more accurate results.

Another approach developed to deal with rare failure events and using a selective sampling approach is presented in [15]. The high-sigma Monte Carlo (HSMC) developed by McConaghy et al. starts by generating a large number of samples, N_{gen}, according to a look-up table based on the target sigma value; as an example, for 6σ analysis 5 billion of samples are generated. Then, a subset of samples is extracted and simulated using an electrical simulator, e.g., 1000 samples. Using the simulation results, a model is trained for each circuit performance, which allows to evaluate the rest of the samples, and based on the simulations order them according to the performance output.

The ordering process permits to identify worst-case points that can be used to accurately predict the distribution tails by electrical simulations. The number of samples simulated in this process is limited by a maximum number of simulations or when the total number of estimated failures samples was simulated. The final yield value is estimated with a 95% confidence interval, considering the total number of

a

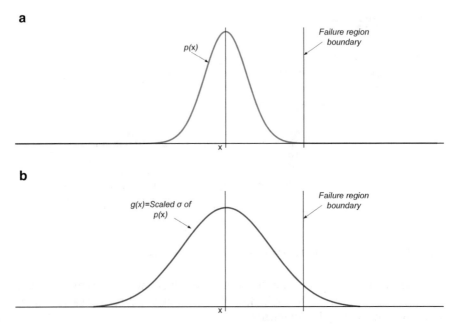

b

Fig. 3.10 Scale-sigma sampling example. (**a**) Original distribution. (**b**) Scaled distribution where the probability of generating samples near the failure region boundary is higher

samples generated initially, and, the number of failed samples, N_{fail}, found by simulating the order worst-case samples, according to:

$$Y_{\text{HSMC}} = \frac{N_{\text{gen}} - N_{\text{fail}}}{N_{\text{gen}}} \qquad (3.24)$$

A similar idea to HSMC, where a large number of samples are considered to comply with circuit performances and are not simulated, was explored by Kuo et al. in [19]. This work presents the importance analysis technique. This technique relies on linear models that relate design variables with circuit performances and permit identifying samples far from performance constraint limits. Then, these samples are excluded from simulations, considering that they are feasible and comply with performance constraints. Using the importance analysis, only samples near the limit performance constraint are simulated. Two circuits were tested using this approach. The first circuit was a phase-locked loop (PLL) and the second a two-stage OpAmp circuit. In the PLL circuit only one performance, the locking time, was considered to define the feasibility constraint and the yield target was solutions with more than 95%. The authors did not provide the computational effort for developing the models used in the tests. The two-stage OpAmp also has one circuit performance measure to define the feasibility and estimate the yield, which was the gain bandwidth (GBW), and the target yield was above 97.5%. The OpAmp circuit results show that from 2000 samples only 213 were simulated when using MCS, and from 400 samples generated by LHS only 50 were simulated.

IS techniques are able to reduce the computational effort of performing a large number of electrical simulations, in the order millions to just a few thousand simulations, but the number of required simulations and the effort to identify failure regions is still very high to include these type of approaches in an optimization loop, which is why these techniques are used, mainly to estimate the yield of selected solutions at the end of the optimization process.

3.1.3.3   Monte Carlo Model-Based Methods

The use of models that relate design variables with circuit performances, allowing replacing the electrical simulator, is another technique used to fast estimate the yield of potential solutions.

In [20] Okobiah et al. presented a yield-aware sizing approach using Kriging metamodels to estimate circuit performances and yield. The Kriging models were created using MATLAB©, and for each of circuit performance a Kriging model was created. The approach was tested using a PLL circuit in a 180 nm technology node, where a total of 21 design variables and process parameters were considered to model circuit performances. The yield-aware sizing approach was compared to typical MC analysis using 1000 samples. The author referred that the Kriging model generation and analysis took a *few hours*, whereas the MC analysis took 5 days to conclude the task; the authors did not specify characteristics of computer running the simulations.

Felt et al., in [21], tested linear and quadratic models to estimate circuit performances under variability. Before generating the models, PCA is used to reduce the number of transistor process parameters, leaving 2–3 statistical relevant process parameters per transistor, which can explain at least 75% of the observed variability.

The models were developed using response surface methodology (RSM) [22]. The coefficients for the RSM models were computed by linear regression, using samples where each parameter is perturbed around the nominal point. The approach was tested to characterize a folded cascode OpAmp, Fig. 3.11, and a PLL circuit. As expected, the quadratic model revealed higher accuracy than the linear model, but since linear models' accuracy is close to the quadratic models result, the authors choose to adopt linear models for the statistical calculations due to the fast implementation.

The use of models allows fast evaluation of a large number of samples, being the main drawback its reusability and the effort to develop accurate models.

3.1.3.4   Non-Monte Carlo Methods

Several works implement yield estimation techniques not based on MC analysis. Although the techniques developed try avoiding a high number of expensive MC

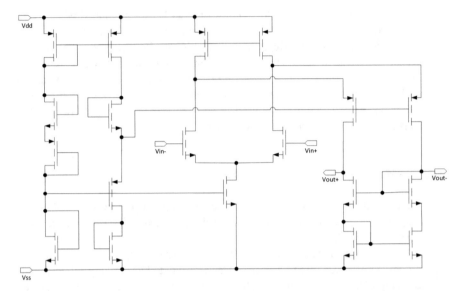

Fig. 3.11 OpAmp schematics adopted by Felt et al. to test their model approach [21]

simulations to estimate the yield, the fact is that most of the presented techniques still depend on a high number of electrical simulations.

The yield estimation technique presented in [23] is an example of this idea. In this work the authors generate models to fast estimate circuit performances and the same models are used to estimate the probability density function of performances, which allows to compute the yield using (3.1). The models are generated and tuned by means of a high number of electrical circuit simulations. So, the yield is not directly estimated by MC analysis, but a high number of simulations are required to develop the yield estimation technique.

Two modelling techniques were tested, the sparse regression (SR) technique and Bayesian model fusion (BMF). The SR technique is used to model circuit performances and yield estimation. For SR modelling technique, consider a performance model as:

$$f(\Delta y) = \sum_{i=1}^{K} a_i g_i(\Delta y) \tag{3.25}$$

where f is circuit performances and $f(\Delta y)$ is a local perturbation around the nominal point, a_i is model coefficients, and $g_i(\Delta y)$ is one of the K basis functions, e.g., linear or quadratic polynomials. By sampling around the nominal point, $\Delta y^{(l)}$ with $l = 1$, $\ldots$, L, where L is the number of samples, it is possible by simulation to evaluate the performance at each sample point, $f^{(l)}$, $l = 1, \ldots, L$, and build a system of linear equations as:

$$f = G \cdot a \tag{3.26}$$

where:

$$f = \left[f^{(1)} \cdots f^{(L)} \right]^{T} \tag{3.27}$$

$$G = \begin{bmatrix} g_1\left(\Delta y^{(1)}\right) & \cdots & g_K\left(\Delta y^{(1)}\right) \\ \vdots & \ddots & \vdots \\ g_1\left(\Delta y^{(L)}\right) & \cdots & g_K\left(\Delta y^{(L)}\right) \end{bmatrix} \tag{3.28}$$

$$a = \left[a_i \cdots a_K \right]^{T} \tag{3.29}$$

Most modelling techniques assume that the number of samples is larger than the number of coefficients or at least equal. Trying to minimize the number of samples may result in fewer samples than coefficients; in this case the solution a is not unique. To solve this problem, SR adds an additional constraint. Considering that only a few of the coefficients are relevant to estimate the circuit performance, the rest of coefficients can be set to zero. So, to find the linear coefficients using SR the following optimization problem must be solved:

$$\min \|G \cdot a - f\|_2^2 \tag{3.30}$$

s.t.

$$\|a\|_0 \leq \lambda \tag{3.31}$$

where $\|\cdot\|_0$ and $\|\cdot\|_2$ denotes the L_0-norm and L_2-norm, respectively, and λ is a regularization parameter of the trade-off between the sparsity of vector a and the value of the cost function being minimized. The L_0-norm of a vector corresponds to the number of non-zeros components of the vector.

The authors instead of adopting the L_0-norm, as described above, implemented the SR technique using the L_1-norm to improve efficiency in solving the optimization problem in (3.30) subject to (3.31). The developed SR models are used by the authors to compute circuit performances and, also, to estimate the probability density function of each performance. Once the performances pdf is known, it is possible to calculate the yield. This approach was tested in an LNA circuit, where the optimization goal was finding the design variables that minimize the power consumption and provide a solution with a yield *sufficiently high* (the term *sufficiently high* was not quantified by the authors). The authors claimed that starting from a solution with less than 50% of yield was possible to improve to a robust solution with performance margin enough to meet the desired yield specification.

The second modelling technique, i.e., BMF technique, was adopted to reduce the modelling cost. This technique was able to take advantage of simulations performed in early stages of the IC design flow, thus reducing the total simulation time to determine the model parameters. The authors referred that were unable to provide details about the BMF technique due to page limitation constraints. But, provide a comparison between SR and BMF using a SRAM bit cell circuit, showing that the BMF technique achieves a $8\times$ times speedup when compared to SR.

An alternative way to understanding the yield is considering a geometric approach. The geometric yield analysis is based on the location of the worst-case statistical parameter vector among all parameter vectors in the parameter feasibility region defined by the circuit performance requirements. To clarify this approach, consider a two-dimensional statistical parameter space with the nominal statistical parameter vector, $\mathbf{x}_{s0}$, and with statistical parameters normally distributed. Also, a performance specification constraint $p > p_L$ that defines a feasibility region R_{sL}, where p_L denotes the lower bound of circuit performance p. Then, the worst-case parameter vector $\mathbf{x}_{sWL}$, for a nominal vector inside the feasibility region, is the vector closest to $\mathbf{x}_{s0}$ among all vectors located on the equal probability isocontour and on the border or outside of the feasibility region in $\overline{R}_{sL}$, (3.32). If the nominal vector is outside the feasibility region, then the worst-case vector is located inside the feasibility region or on the border, (3.33). The subscript WL denotes that the vector is a worst-case parameter with respect to a lower bound. The upper bound worst-case vector, WU, is defined in a similar manner considering a feasibility region defined by an upper bound performance constraint.

$$\mathbf{x}_{s0} \in R_{sL/U} : \quad \max_{\mathbf{x}_s} \ \text{pdf}(\mathbf{x}_s) \quad \text{s.t.} \ \mathbf{x}_s \in \overline{R}_{sL/U} \qquad (3.32)$$

$$\mathbf{x}_{s0} \in \overline{R}_{sL/U} : \quad \max_{\mathbf{x}_s} \ \text{pdf}(\mathbf{x}_s) \quad \text{s.t.} \ \mathbf{x}_s \in R_{sL/U} \qquad (3.33)$$

The normal pdf() defines curves of constant value as previously presented in (2.49). The relation between the pdf() value and $\beta^2(\mathbf{x}_s)$ shows that an increase in the value of the pdf() function enforces a decrease in $\beta^2(\mathbf{x}_s)$, since a smaller ellipsoid is closer to the nominal vector where the distribution is centered. Using this relation, it is possible to express the worst-case vector defined in (3.32) and (3.33) in terms of β, respectively, as in (3.34) and (3.35).

$$\mathbf{x}_{s0} \in R_{sL/U} : \quad \min_{\mathbf{x}_s} \ \beta^2(\mathbf{x}_s) \quad \text{s.t.} \ \mathbf{x}_s \in \overline{R}_{sL/U} \qquad (3.34)$$

$$\mathbf{x}_{s0} \in \overline{R}_{sL/U} : \quad \min_{\mathbf{x}_s} \ \beta^2(\mathbf{x}_s) \quad \text{s.t.} \ \mathbf{x}_s \in R_{sL/U} \qquad (3.35)$$

The constraint in (3.34) and (3.35) can be expressed using the upper and lower bounds of the performance, and, also, considering the range parameters $\mathbf{x}_r$ that define a box range region R_r, and, also constraint the feasibility region. Detailing the constraints results in new expressions to define the worst-case vector, (3.36)–(3.39), separating the upper and lower bound and according to the position of the nominal vector

$$p \geq p_L \wedge \mathbf{x}_{s0} \in R_{sL} : \quad \min_{\mathbf{x}_s} \ \beta^2(\mathbf{x}_s) \quad \text{s.t.} \quad \min_{\mathbf{x}_r \in R_r} p(\mathbf{x}_s, \mathbf{x}_r) \leq p_L \quad (3.36)$$

$$p \geq p_L \wedge \mathbf{x}_{s0} \in \overline{R}_{sL} : \quad \min_{\mathbf{x}_s} \ \beta^2(\mathbf{x}_s) \quad \text{s.t.} \quad \min_{\mathbf{x}_r \in R_r} p(\mathbf{x}_s, \mathbf{x}_r) \geq p_L \quad (3.37)$$

$$p \leq p_U \wedge \mathbf{x}_{s0} \in R_{sU} : \quad \min_{\mathbf{x}_s} \ \beta^2(\mathbf{x}_s) \quad \text{s.t.} \quad \max_{\mathbf{x}_r \in R_r} p(\mathbf{x}_s, \mathbf{x}_r) \geq p_U \quad (3.38)$$

$$p \leq p_U \wedge \mathbf{x}_{s0} \in \overline{R}_{sU} : \quad \min_{\mathbf{x}_s} \ \beta^2(\mathbf{x}_s) \quad \text{s.t.} \quad \max_{\mathbf{x}_r \in R_r} p(\mathbf{x}_s, \mathbf{x}_r) \leq p_U \quad (3.39)$$

The solution of the optimization problems defined in (3.36)–(3.39), according to [1], is given by:

$$p \geq p_L \wedge \mathbf{x}_{s0} \in R_{sL} \quad \text{and} \quad p \leq p_U \wedge \mathbf{x}_{s0} \in \overline{R}_{sU}$$

$$\rightarrow \mathbf{x}_{sWL/U} - \mathbf{x}_{s0} = \frac{-\beta_{WL/U}}{\sqrt{\nabla p(\mathbf{x}_{sWL/U})^T \mathbf{C} \cdot \nabla p(\mathbf{x}_{sWL/U})}} \cdot \mathbf{C} \cdot \nabla p(\mathbf{x}_{sWL/U}) \quad (3.40)$$

$$p \geq p_L \wedge \mathbf{x}_{s0} \in \overline{R}_{sU} \quad \text{and} \quad p \leq p_U \wedge \mathbf{x}_{s0} \in R_{sL}$$

$$\rightarrow \mathbf{x}_{sWL/U} - \mathbf{x}_{s0} = \frac{\beta_{WL/U}}{\sqrt{\nabla p(\mathbf{x}_{sWL/U})^T \cdot \mathbf{C} \cdot \nabla p(\mathbf{x}_{sWL/U})}} \cdot \mathbf{C} \cdot \nabla p(\mathbf{x}_{sWL/U}) \quad (3.41)$$

where $\beta_{WL/U}$ represents the worst-case distance, which defines the ellipsoid that touches the boundary of the feasibility region, and is expressed as:

$$\beta_{WL/U}^2 = \beta^2(\mathbf{x}_{sWL/U}) = (\mathbf{x}_{sWL/U} - \mathbf{x}_{s0})^T \mathbf{C}^{-1} (\mathbf{x}_{sWL/U} - \mathbf{x}_{s0}) \quad (3.42)$$

Using the linearization of performance p at the worst-case parameter vector, $\overline{p}_{WLU}(\mathbf{x}_s) = p_{L/U} + \nabla p(\mathbf{x}_{sWL/U})^T \cdot (\mathbf{x}_s - \mathbf{x}_{sWL/U})$ and considering the nominal vector $- \nabla p(\mathbf{x}_{sWL/U})^T \cdot (\mathbf{x}_{sWL/U} - \mathbf{x}_{s0}) = \overline{p}_{WLU}(\mathbf{x}_{s0}) - p_{L/U}$, it is possible to obtain the worst-case distance from (3.40) and (3.41):

$$p \geq p_L \wedge \mathbf{x}_{s0} \in R_{sL} \quad \text{and} \quad p \leq p_U \wedge \mathbf{x}_{s0} \in \overline{R}_{sU} \rightarrow \beta_{WL/U}$$

$$= \frac{\overline{p}_{WLU}(\mathbf{x}_{s0}) - p_{L/U}}{\sqrt{\nabla p(\mathbf{x}_{sWL/U})^T \cdot \mathbf{C} \cdot \nabla p(\mathbf{x}_{sWL/U})}} = \frac{\nabla p(\mathbf{x}_{sWL/U})^T \cdot (\mathbf{x}_{s0} - \mathbf{x}_{sWL/U})}{\sigma_{\overline{p}_{WLU}}} \quad (3.43)$$

$$p \geq p_L \wedge \mathbf{x}_{s0} \in \overline{R}_{sU} \quad \text{and} \quad p \leq p_U \wedge \mathbf{x}_{s0} \in R_{sL} \rightarrow \beta_{WL/U}$$

$$= \frac{p_{L/U} - \overline{p}_{WLU}(\mathbf{x}_{s0})}{\sqrt{\nabla p(\mathbf{x}_{sWL/U})^T \cdot \mathbf{C} \cdot \nabla p(\mathbf{x}_{sWL/U})}} = \frac{\nabla p(\mathbf{x}_{sWL/U})^T \cdot (\mathbf{x}_{sWL/U} - \mathbf{x}_{s0})}{\sigma_{\overline{p}_{WLU}}} \quad (3.44)$$

where $\sigma_{\overline{p}_{WLU}}$ is the standard deviation of the linearized performance at the worst-case parameter vector.

Considering the worst-case distance based on the linearized performance and a normal distribution with zero mean and unit variance, the yield for a circuit performance according to the feature bound, can be estimated by (3.45) or (3.46).

$$Y_U = \begin{cases} \int_{-\infty}^{\beta_{WU}} \frac{1}{\sqrt{2\pi}} e^{-t^2/2} dt, & p(\mathbf{x}_{s0}) \leq p_U \\ \int_{-\infty}^{-\beta_{WU}} \frac{1}{\sqrt{2\pi}} e^{-t^2/2} dt, & p(\mathbf{x}_{s0}) \geq p_U \end{cases} \quad (3.45)$$

$$Y_L = \begin{cases} \int_{-\infty}^{\beta_{WL}} \frac{1}{\sqrt{2\pi}} e^{-t^2/2} dt, & p(\mathbf{x}_{s0}) \geq p_L \\ \int_{-\infty}^{-\beta_{WL}} \frac{1}{\sqrt{2\pi}} e^{-t^2/2} dt, & p(\mathbf{x}_{s0}) \leq p_L \end{cases} \quad (3.46)$$

Since β_W is a multiple of sigma of the normal distribution, the yield values with respect to a circuit performance can be obtained from statistical tables, like Table 2.4. In terms of accuracy, this approach presents an absolute yield error around 1–3%, which decreases for higher yield values [1]. Also, a closer circuit performance model bound, in the presented example was adopted a linear model, to the real unknown feasibility border helps decreasing the error, which is why some approaches adopt quadratic models. To create the linear model, in addition to the nominal vector simulation, to evaluate the circuit performances at this point, are required n simulations more per iteration, where n corresponds to the total number of the circuit parameter and process variables. As an example, consider a circuit having 20 devices and a total of 13 variables per device. So, to build the model it would be necessary to perform $1 + 20 \times 13 = 261$ simulations per iteration of the optimization algorithm. Building a quadratic model requires a much large number of simulations, according to [15] $1 + n(n - 1)/2$ simulations are required. So, using the previous example values, the number of simulations per iteration is: $1 + 20 \times 13 \times (20 \times 13 - 1)/2 = 33{,}671$. Adopting the linear model has the advantage of being a very fast method to estimate the yield for small circuits and with few variables; the major drawback is the accuracy of the estimation due to modelling the true performance bound by a linear model. The quadratic model may not be accurate enough for many circuits [15], with the overhead of requiring a larger number of simulations than the linear model.

From the yield expressions (3.45) and (3.46) results that the desired increase in the yield value has to come from an increase in the worst-case distance β_W. There are two ways to increase the worst-case distance, the first is to decrease the standard deviation of the linearized performance function, which is the same as decreasing the sensitivity of the circuit performance with respect to the statistical parameters. The second approach, which is adopted by design centering techniques, is to increase the distance between the nominal parameter vector and the worst-case parameter vector for a nominal parameter vector inside the feasibility region or reducing the distance if the nominal parameter vector is outside the feasibility region. Including a linear term, related to the deterministic circuit design variables, $\mathbf{x}_d$, in (3.43) and (3.44) allows developing a technique to optimize the yield by finding the best combination of design variables and statistical parameters using the worst-case distance (WCD) [1].

$$p \geq p_L \wedge \mathbf{x}_{s0} \in R_{sL} \quad \text{and} \quad p \leq p_U \wedge \mathbf{x}_{s0} \in \bar{R}_{sU} \rightarrow$$

$$\rightarrow \beta_{WL/U} = \frac{\nabla p(\mathbf{x}_{sWL/U})^T \cdot (\mathbf{x}_{s0} - \mathbf{x}_{sWL/U}) + \nabla p(\mathbf{x}_{d\mu})^T \cdot (\mathbf{x}_d - \mathbf{x}_{d\mu})}{\sigma_{\bar{p}_{WLU}}} \quad (3.47)$$

$$p \geq p_L \wedge \mathbf{x}_{s0} \in \bar{R}_{sU} \quad \text{and} \quad p \leq p_U \wedge \mathbf{x}_{s0} \in R_{sL} \rightarrow$$

$$\rightarrow \beta_{WL/U} = \frac{\nabla p(\mathbf{x}_{sWL/U})^T \cdot (\mathbf{x}_{sWL/U} - \mathbf{x}_{s0}) - \nabla p(\mathbf{x}_{d\mu})^T \cdot (\mathbf{x}_d - \mathbf{x}_{d\mu})}{\sigma_{\bar{p}_{WLU}}} \quad (3.48)$$

The main idea of design centering is finding a robust solution by maximizing the inscribed ellipsoid in the feasibility region in design space, Fig. 3.12.

Fig. 3.12 Design centering in two-dimensional design space

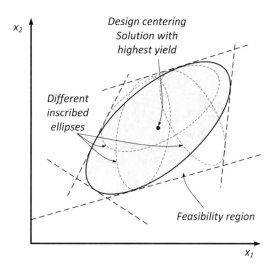

Due to the relation between design space and performance space several works implement worst-case performance (WCP) techniques to optimize the yield. In [24], the authors presented a technique called Mesh adaptive direct search (MADS) to solve the problem of finding the WCD and WCP to estimate the yield of the circuit solutions. The approach was tested in two different amplifier topologies and the results were compared to a commercial tool that implements SQP to find the WCD. The test shows similar results between MADS and the commercial tools, with the authors claiming that MADS performs best on noisy nonlinear problems, where SQP was unable to find a solution.

Pan and Graeb [25] used the WCD-based approach to estimate the yield under variability condition and also consider the aging effects on the circuit, which the authors called lifetime yield. The WCD approach was tested on a two-stage Miller OTA, comparing the use of a linear model and a quadratic model. The results show for the linear model an average error of 8.5% in accuracy when compared to the WCD obtained from a commercial tool, whereas the quadratic model presents an average error of 4.8%. As expected, the linear model presents better execution time than the quadratic model.

In [26] fuzzy set theory was incorporated into WCD linear performance models to define a possibility measure of performance failure, which allows to improve the yield by minimizing the possibility of failure. The new approach was tested on a two-stage OTA, Fig. 3.13, where only four process parameters and two environment parameters were considered as sources of variability.

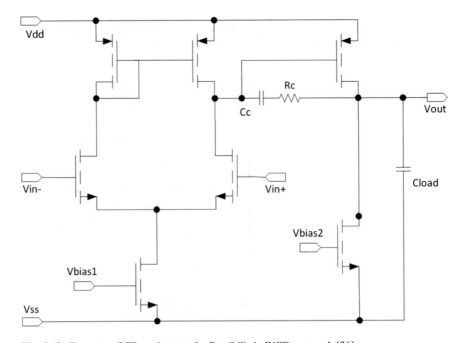

Fig. 3.13 Two-stage OTS used to test the Possibilistic-WCD approach [26]

The final sized OTA circuit was compared with an approach were an overdesign methodology was adopted. The results show that the Possibilistic-WCD approach was able to achieve solutions with higher yield values than the overdesign methodology.

Opalski in [27] presents a study on the computational effort of design centering methods. The author concluded that WCD-based approaches are an alternative for a MC-based approach in problems with a small number of variability sources, and that for larger circuits with a higher number of variables the computational cost could easily exceed the corresponding computational cost of MC-based approaches. This conclusion is shared by other researchers, pointing that the increase of process variables shows heavy degradation in execution time [15]. Opalski also concluded that statistical design considering variability effects typically adopts techniques with large potential for parallelization, thus making possible to reduce the overall execution time and creating statistical design techniques with a computational execution time comparable to nominal design on a single CPU.

Before finishing this section, it is important to refer the layout parasitic effects, to which analog ICs are quite sensitive and can cause severe performance degradation, thus affecting the circuit final yield value. Estimating the impact of the parasitic effects in the circuit is only possible after having a sized circuit solution and a complete physical layout design of the circuit. Performing, during an optimization circuit sizing process, a complete automatic layout for each potential solution in order to predict the impact of parasitics on each potential solution is still a very difficult and computationally expensive task. In [28] an automatic layout generation based on templates with parasitic extraction inside the optimization loop is presented. This approach presents a good solution to include parasitics in the sizing process but is very dependent on the adopted technology and the circuit being optimized. To overcome this difficulty, most works that incorporate yield estimation with parasitic effects in the circuit sizing and optimization process start from a feasible circuit solution, close to the final circuit, where a layout is provided and an initial estimation for parasitic effects already exists [20, 29, 30]. In [30], the authors presented a method called Design-Of-Experiments assisted Monte Carlo (DOE-MC), where several MC simulations are performed to create statistical modelling for circuit performances, which are later used during the circuit optimization. In their approach the layout structure is static, otherwise the layout design could change significantly causing different circuit parasitics than the ones considered during the initial sizing and optimization process. This type of approach is not possible to implement in automatic circuit sizing tools, where the search for a fully sized circuit (s) solution(s) is achieved by solution space exploration. A fully automatic analog IC synthesis tool including yield-aware, layout-aware, and parasitic-aware features will only be possible when the layout and routing generation process for analog ICs is completely automated, like in its digital counterpart.

3.2   Commercial EDA tools

The growing impact of variability effects in new nanometer technology nodes created new business opportunities for the EDA industry. Nowadays, different commercial EDA tools compete to develop and offer to the analog IC designers' community the most reliable yield estimation technique, and, also, computational efficient yield-aware circuit sizing and optimization processes. Most of those techniques derived from state-of-the-art academic works are presented in Sect. 3.1. For example, Solido Design Automation implements the high-sigma Monte Carlo [31] developed by McConaghy et al., for lower yield requirements, up to 4σ, this company offers a tool named Fast Monte Carlo [32] which basically interrupts the MC simulations if the target yield value was achieved. Whereas, MunEDA [33] offers worst-case analysis techniques as solution to predict the yield for high-sigma requirements, e.g., 6σ to 9σ, while for not so demanding yield solutions, 3σ to 5σ, adopts MC analysis, thus trading efficiency for accuracy. Synopsys, Inc. [34] adopts MC analysis using LHS or QMC to reduce the number of samples required for yield estimation, and also in their reliability tool a technique is implemented called *Sigma Amplification* to reduce the number of samples for circuits with 3σ to 4.5σ yield requirements. Although Synopsys do not reveal many details, this technique is, probably, based on the scaled-sigma sampling technique presented earlier. Another major player in the EDA industry is Cadence. The Virtuoso Analog Design Environment (ADE) XL [35] from Cadence offers typical MC analysis using random, LHS, and LDS sampling techniques for yield estimation. The GXL version of ADE implements the WCD approach for high-sigma yield requirements. Also, at the ADE-GXL tool, the scale-sigma sampling technique for high-sigma yield estimation is implemented [36].

3.3   Conclusion

Although there are many different techniques to estimate the yield, MC is still considered the gold standard for yield estimation. According to Cadence® [37], MC analysis is the solution for accurately estimating the yield and capturing the statistics on circuit behavior in advanced integration nodes. From the analysis of the commercial EDA tools is notorious the importance given to yield-aware and variation-techniques, but the analog IC design flow in most of those tools is still very dependent on human intervention.

In Table 3.4, a summary of the yield estimation techniques adopted by commercial EDA tools is presented. Also, in Table 3.5, a summary of the different yield estimation techniques described in this chapter is presented.

Table 3.4 Yield estimation techniques in commercial EDA tools

Company	Yield estimation technique according to yield targets			
	Up to 5σ		Above 5σ	
Solido	Fast Monte Carlo	[32]	High-sigma MC	[31]
MunEDA	Typical MC analysis	[33]	Worst-case analysis	[33]
Synopsys	MC analysis using LHS or QMC Sigma amplification	[34]		
Cadence	Typical MC analysis MC analysis using LHS or QMC	[35]	Worst-case analysis Scale-sigma sampling	[36]

Table 3.5 Yield estimation approaches

	Monte Carlo-based			Non-Monte Carlo-based
	MC/Quasi-MC	Importance Sampling Selective sampling	Model-based	
Works	Singhee [2] Liu [3] Guerra-Gomez [5] Afacan [7, 8, 10] Pak [11]	Qazi [12] Yilmaz [13] Kanj [14] McConaghy [15] Yao [16] Wang [17] Sun [18] Kuo [19]	Kuo [19] Okobiah [20] Felt [21]	Graeb [1] WCD/WCP Li [23] Bűrmen [24] WCD/WCP Pan [25] WCD Sciacca [26] Fuzzy-WCD Opalski [27] WCD
Strengths	Accuracy. Can handle a large number of input process variables.	High-sigma analysis. Explores failure region near the distribution tails.	Fast yield estimation	High-sigma analysis. Fast yield estimation.
Weaknesses	Time consuming.	Distorts the sampling distribution. Limited input process variables.	Accuracy. Reusability of the models.	Accuracy. Limited input process variables.
Strategy	QMC/LHS evenly sample parameter space. or Two-step MC analysis to identify promising solutions.	Samples the space around the failure region. or Selects regions of interest to perform sampling.	Uses models for fast MC simulations.	Performs at least one additional simulation per parameter variable to create a model for WC. Yield is indirectly estimated based on multiples of σ.

References

1. H.E. Graeb, *Analog Design Centering and Sizing* (Springer, Dordrecht, 2007)
2. A. Singhee, R.A. Rutenbar, Why quasi-Monte Carlo is better than Monte Carlo or Latin hypercube sampling for statistical circuit analysis. IEEE Trans. Comput. Aided Des. Integr. Circuits Syst. **29**(11), 1763–1776 (2010)
3. B. Liu, F.V. Fernandez, G.G.E. Gielen, Efficient and accurate statistical analog yield optimization and variation-aware circuit sizing based on computational intelligence techniques. IEEE Trans. Comput. Aided Des. Integr. Circuits Syst. **30**(6), 793–805 (2011)
4. Y.-C. Ho, An explanation of ordinal optimization: Soft computing for hard problems. Inf. Sci. **113**(3–4), 169–192 (1999)
5. I. Guerra-Gomez, E. Tlelo-Cuautle, L.G. de la Fraga, OCBA in the yield optimization of analog integrated circuits by evolutionary algorithms, in *2015 IEEE Int. Symp. Circuits Syst. (ISCAS)*, Lisbon, 2015
6. K. Deb, R. Agrawal, Simulated binary crossover for continuous search space. Complex Syst. **9** (2), 115–148 (1995)
7. E. Afacan, G. Berkol, A.E. Pusane, G. Dündar, F. Başkaya, Adaptive sized Quasi-Monte Carlo based yield aware analog circuit optimization tool, in *2014 5th Eur. Workshop on CMOS Variability (VARI)*, Palma de Mallorca, 2014
8. E. Afacan, G. Berkol, A.E. Pusane, G. Dündar, F. Başkaya, A hybrid Quasi Monte Carlo method for yield aware analog circuit sizing tool, in *2015 Des. Automat. Test Eur. Conf. Exhibition (DATE)*, Grenoble, 2015
9. H. Niederreiter, Constructions of (t,m,s)-nets and (t,s)-sequences. Finite Fields Appl. **11**(3), 578–600 (2005)
10. E. Afacan, G. Berkol, G. Dundar, A.E. Pusane, F. Baskaya, An analog circuit synthesis tool based on efficient and reliable yield estimation. Microelectron. J. **54**, 14–22 (2016)
11. M. Pak, F.V. Fernandez, G. Dundar, Comparison of QMC-based yield-aware pareto front techniques for multi-objective robust analog synthesis. Integration VLSI J. **55**, 357–365 (2016)
12. M. Qazi, M. Tikekar, L. Dolecek, D. Shah, A. Chandrakasan, Loop flattening & spherical sampling: highly efficient model reduction techniques for SRAM yield analysis, in *2010 Des. Autom. Test Eur. Conf. Exhibition (DATE)*, Dresden, 2010
13. E. Yilmaz, S. Ozev, Adaptive-learning-based importance sampling for analog circuit DPPM estimation. IEEE Des. Test **32**(1), 36–43 (2015)
14. R. Kanj, R. Joshi, S. Nassif, Mixture importance sampling and its application to the analysis of SRAM designs in the presence of rare failure events, in *2006 43rd ACM/IEEE Design Automat. Conf.*, San Francisco, CA, 2006
15. T. McConaghy, K. Breen, J. Dyck, A. Gupta, *Variation-Aware Design of Custom Integrated Circuits: A Hands-on Field Guide* (Springer, New York, 2013)
16. J. Yao, Z. Ye, Y. Wang, Importance boundary sampling for SRAM yield analysis with multiple failure regions. IEEE Trans. Comput. Aided Des. Integr. Syst. **33**(3), 384–396 (2014)
17. M. Wang, C. Yan, X. Li, D. Zhou, X. Zeng, High-dimensional and multiple-failure-region importance sampling for SRAM yield analysis. IEEE Trans. Very Large Scale Integr. (VLSI) Syst. **25**(3), 806–819 (2017)
18. S. Sun, X. Li, H. Liu, K. Luo, B. Gu, Fast statistical analysis of rare circuit failure events via scaled-sigma sampling for high-dimensional variation space. IEEE Trans. Comput. Aided Des. Integr. Circuits Syst. **34**(7), 1096–1109 (2015)
19. C.C. Kuo, W.Y. Hu, Y.H. Chen, J.F. Kuan, Y.K. Cheng, Efficient trimmed-sample Monte Carlo methodology and yield-aware design flow for analog circuits, in *DAC Des. Autom. Conf.*, San Francisco, CA, 2012
20. O. Okobiah, S.P. Mohanty, E. Kougianos, Fast statistical process variation analysis using universal Kriging metamodeling: a PLL example, in *2013 IEEE 56th Int. Midwest Symp. Circuits Syst. (MWSCAS)*, Columbus, OH, 2013

21. E. Felt, S. Zanella, C. Guardiani, A. Sangiovanni-Vincentelli, Hierarchical statistical charac-terization of mixed-signal circuits using, in *Proc. Int. Conf. Comput. Aided Des.*, San Jose, CA, 1996
22. R.H. Myers, D.C. Montgomery, C.M. Anderson-Cook, *Response Surface Methodology: Process and Product Optimization Using Designed Experiments*, Wiley Series in Probability and Statistics, 4th edn. (Wiley Inc. Publications, Hoboken, NJ, 2016)
23. X. Li, W. Zhang, F. Wang, Large-scale statistical performance modeling of analog and mixed-signal circuits, in *Proc. IEEE 2012 Custom Integr. Circuits Conf.*, San Jose, CA, 2012
24. Á. Bűrmen, H. Habal, Computing worst-case performance and yield of analog integrated circuits by means of mesh adaptive direct search. J. Microelectron. Electron. Comp. Mater. **45**(2), 160–170 (2015)
25. X. Pan, H. Graeb, Reliability analysis of analog circuits using quadratic lifetime worst-case distance prediction, in *IEEE Custom Integrated Circuits Conf. 2010*, San Jose, CA, 2010
26. E. Sciacca, S. Spinella, A.M. Anile, Possibilistic worst case distance and applications to circuit sizing, in *Theoretical Advances and Applications of Fuzzy Logic and Soft Computing*, (Springer, Berlin, 2007), pp. 287–295
27. L. Opalski, Remarks on statistical design centering. Int. J. Electron. Telecommun. **57**(2), 159–167 (2011)
28. R. Martins, N. Lourenço, A. Canelas, N. Horta, Stochastic-based placement template generator for analog IC layout-aware synthesis. Integration VLSI J. **58**, 485–495 (2017)
29. D. Ghai, S.P. Mohanty, E. Kougianos, Design of parasitic and process-variation aware nano-CMOS RF circuits: a VCO case study. IEEE Trans. Very Large Scale Integr. (VLSI) Syst. **17**(9), 1339–1342 (2009)
30. S.P. Mohanty, E. Kougianos, Incorporating manufacturing process variation awareness in fast design optimization of nanoscale CMOS VCOs. IEEE Trans. Semicond. Manuf. **27**(1), 22–31 (2014)
31. Solido Design Automation, Mentor a Siemens Business, Technology|Solido Design Automation High Sigma Monte Carlo. [Online]. Available: https://www.solidodesign.com/products/variation-designer/technology/#high-sigma-monte-carlo. Accessed 6 Nov 2018
32. Solido Design Automation, Mentor a Siemens Business, Technology|Solido Design Automation Fast Monte Carlo. [Online]. Available: https://www.solidodesign.com/products/variation-designer/technology/#fast-monte-carlo. Accessed 8 Nov 2018
33. MunEDA GmbH, muneda.com—Solutions Analysis Overview, 2018. [Online]. Available: https://www.muneda.com/solutions_analysis_overview.php. Accessed 9 Nov 2018
34. Synopsys, Inc., Reliability Analysis. [Online]. Available: https://www.synopsys.com/verification/ams-verification/reliability-analysis.html. Accessed 9 Nov 2018
35. Cadence Design Systems, Inc., Virtuoso Analog Design Environment Family, 2014. [Online]. Available: https://www.cadence.com/content/dam/cadence-www/global/en_US/documents/tools/custom-ic-analog-rf-design/virtuoso-analog-design-fam-ds.pdf. Accessed 13 Sept 2018
36. Cadence Design Systems, Inc., Vituoso Variation Option. [Online]. Available: https://www.cadence.com/content/cadence-www/global/en_US/home/tools/custom-ic-analog-rf-design/circuit-design/virtuoso-variation-option.html. Accessed 9 Nov 2018
37. Cadence Design Systems, Inc., Accelerating Monte Carlo Analysis at Advanced Nodes, 2016. [Online]. Available: https://www.cadence.com/content/dam/cadence-www/global/en_US/documents/tools/custom-ic-analog-rf-design/monte-carlo-analysis-at-advanced-nodes-wp.pdf. Accessed 20 Jan 2018

Chapter 4
Monte Carlo-Based Yield Estimation: New Methodology

4.1 New MC-Based Yield Estimation Methodology: General Description

The main goals of this work were pointed at Chap. 1, being the most challenging objective the adoption of MC analysis for yield estimation in a population-based optimization algorithm. The main drawback of simulation-based MC approaches, in order to estimate a circuit yield, is the heavy computational burden it represents, especially in the new IC sizing and optimization tools that implements evolutionary-based optimization techniques, where a large number of potential solutions have to be evaluated inside the optimization loop. The time impact of performing MC simulations in an evolutionary-based optimization algorithm can easily be estimated using (4.1).

$$T_{MC} = S_T \times P \times G \times S_{MC} \qquad (4.1)$$

where

S_T—Simulation time per potential solution;
P—Number of individuals in the population;
G—Number of generations;
S_{MC}—Number of MC simulations/iterations performed on each potential solution.

Setting the previous parameters to typical values, used in the circuits presented at the results chapter, $P = 256$, $G = 300$, $S_{MC} = 500$ and the simulation time to an optimistic value $S_T = 0.02$ s. The total simulation time added to the optimization process for yield estimation is $T_{MC} = 768,000$ s, which corresponds to 213 h or almost 9 days.

Since MC analysis for yield estimation is largely accepted and adopted by IC designers' and the IC industry, two techniques have been used to reduce the impact of MC simulations; the first reduces the impact by performing a smaller number of

© Springer Nature Switzerland AG 2020   97
A. M. L. Canelas et al., *Yield-Aware Analog IC Design and Optimization in Nanometer-scale Technologies*, https://doi.org/10.1007/978-3-030-41536-5_4

simulations to each potential solution, like QMC; the second selects a reduced
number of potential solutions to perform MC analysis. The adopted approach in
this thesis implements the second technique, where a larger number of MC simulations is performed on a handful of individuals selected from the population-based
optimization algorithm. In order to keep a good accuracy in the yield estimation,
when using only a small number of elements/individuals, it is necessary to identify
and select relevant elements of the population to submit to MC analysis. The fact that
the population may include potential solutions from very different areas of the
feasibility region leads to an approach where solutions close to each other must be
grouped together and from each group a representative or relevant element should be
selected. Based on the previous idea, a natural choice to perform the grouping of
similar solutions and selection of relevant elements is a clustering algorithm. Clustering allows the reduction of the original data set into a smaller representative subset
[1]. Also, it is possible to achieve data dimensionality reduction [2]. Considering the
cluster data reduction feature, a novel approach for an MC-based yield estimation
technique for population-based optimization algorithms will be developed. The
general flow of this new yield estimation methodology is presented in Fig. 4.1.

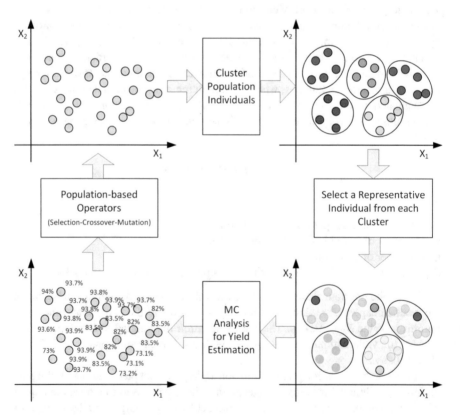

Fig. 4.1 Proposed MC-based yield estimation methodology for population-based circuit sizing and
optimization tools

The new yield estimation methodology at each generation of the population-based optimization algorithm receives as input the population individuals, which are then clustered in the variable space. The next step selects from each cluster a representative individual to perform MC analysis and accurately estimate the yield. Based on the representative individual yield-estimated value, the yield for the rest of the population is estimated, and the output of the new methodology is a completely evaluated population. After population evaluation the population-based optimization algorithm applies the typical operators to produce the next generation of potential solutions. This yield estimation technique allows the inclusion of the yield as an objective of the analog IC sizing and optimization process with a reduced time impact from the MC simulations by modifying the evaluation stage of the population-based algorithm.

Considering (4.1) it is possible to estimate the time impact when using this approach to perform MC-based yield estimation, for that consider setting 5 clusters in each generation, so $P = 5$, and the same values for the rest of the parameters as in the previous example. With these values the total time is $T_{MC} = 15{,}000$ s, around 4 h, which represents a reduction of 98% in simulation time.

Since one of the techniques to support the new yield estimation methodology is clustering, in Sect. 4.2 an overview of several clustering algorithms is presented. Also, the selection of the representative individual from each cluster is discussed, and several problems identified at early tests of the methodology are presented and corrected.

4.2 Clustering Overview

Unsupervised classification or clustering main goal is to identify different and meaningful classes or attributes in unlabeled datasets based on some measure of similitude among the data points. Similarity usually reflects a degree of proximity among the data points [3]. The concept of proximity, in some problems, can be related to a distance measure between data points in some feature space, which inversely relates to similarity, since smaller values in distance represents higher similarity values among data samples [4]. Most popular clustering algorithms are iterative methodologies, which adopts a distance measure in order to group data points around a cluster center by minimizing a cost function that measures the compactness of the data points around the centers [5]. Also, some clustering techniques improve data partitioning and compactness by minimizing the distance between data points in a cluster and maximizing the distance between clusters to improve data separation, Fig. 4.2.

Finding the partitions that result in the optimum clusters definition may seem an easy problem to solve since the number of data points is finite. However, in clustering algorithms, like k-means, it is possible to demonstrate that this is an NP-hard problem [6] and exhaustive search for the optimum solutions is not

Fig. 4.2 Clustering similar
data points by minimizing
intra-cluster distances and
maximizing inter-cluster
distances

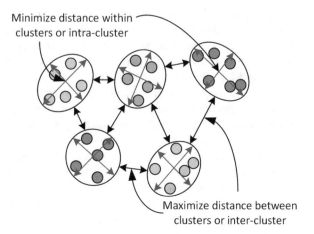

possible, since the number of non-empty partitions for n data points into K disjoint
groups is a Stirling number of the second kind [7], which is given by (4.2).

$$S_n^{(K)} = \frac{1}{K!} \sum_{i=1}^{K} (-1)^{K-i} \binom{K}{i} i^n \qquad (4.2)$$

where $\binom{K}{i} = \frac{K!}{i!(K-i)!}$.

According to [8], using (4.2) for $n = 25$ data points and a number of clusters
$K = 4$, there are approximately 4.69×10^{13} different disjoint partitions. The
impressive number of possible partitions is increased if the problem must also
explore all possible number of clusters combinations, in order to find the optimum
number of clusters to group n data points (4.3):

$$\sum_{k=2}^{n} S_n^{(k)} \qquad (4.3)$$

where $S_n^{(k)}$ is defined as in (4.2).

The clustering techniques detailed in the next section were studied and tested in
order to become part of the new methodology for yield estimation with reduced time
impact from the MC simulations.

4.2.1   K-Means Clustering Algorithm

Among the most popular partitional clustering algorithms is the k-means (KMS) [9]
clustering algorithm. A data set $D = \{\mathbf{x}_1, \ldots, \mathbf{x}_n\}$, with n data points in a d-
dimensional space and an input parameter $k > 1$, is considered. The KMS clustering

algorithm can group the data point into a set of disjoint partitions or clusters C_i, with $i = 1, \ldots, k$, by minimizing the distance between data points in a cluster to its center, where the center is computed as the mean value of the data points in the cluster, which is reached in an iterative optimization process that minimizes the following cost function (4.4):

$$J_{\text{KMS}} = \sum_{i=1}^{k} \sum_{\mathbf{x} \in C_i} \|\mathbf{x} - \mathbf{c}_i\|_2^2 \tag{4.4}$$

where k is the number of clusters, $\mathbf{x} = (x_1, \ldots, x_d)^T$ are data points from a d-dimensional space belonging to cluster C_i, $\| \cdot \|_2^2$ is the squared Euclidian distance, and $\mathbf{c}_i$ is the mean or centroid data point of cluster C_i at one iteration of the KMS algorithm.

Using the indicator function (3.3), it is possible to define the allocation of data point $\mathbf{x}_j = \left(x_j^{(1)}, \ldots, x_j^{(d)} \right)^T$ to cluster C_i and compute each component of cluster I centroid data point $\mathbf{c}_i = \left(c_i^{(1)}, \ldots, c_i^{(d)} \right)^T$ needed by the cost function (4.4) at each iteration of KMS:

$$c_i^{(s)} = \frac{\sum_{j=1}^{n} 1_{C_i}\left(\mathbf{x}_j\right) x_j^{(s)}}{\sum_{j=1}^{n} 1_{C_i}\left(\mathbf{x}_j\right)}; \quad s = 1, \ldots, d \tag{4.5}$$

Based on the previous description, a possible implementation of the KMS algorithm is detailed in Algorithm 4.1. The stopping criterion adopted in Algorithm 4.1 checks if the cost function values between consecutive iterations are below some small value, which may be an indicator that the algorithm has converged. It is also typical to define a maximum number of iterations to avoid infinite loops.

The standard KMS algorithm presented so far only considers the minimization of the within-cluster distance, (4.4). Improving cluster separability is possible by defining a second cost function that measures the between-cluster distance, which must be maximized. The new KMS optimization problem can be presented in the form of scatter matrices [10]:

$$S_{\text{W}} = \sum_{i=1}^{k} \sum_{j=1}^{n} 1_{C_i}\left(\mathbf{x}_j\right)\left(\mathbf{x}_j - \mathbf{c}_i\right)\left(\mathbf{x}_j - \mathbf{c}_i\right)^T \tag{4.6}$$

$$S_{\text{B}} = \sum_{i=1}^{k} \left(\mathbf{c}_i - \mathbf{C}\right)\left(\mathbf{c}_i - \mathbf{C}\right)^T \quad \text{where} \quad \mathbf{C} = \frac{1}{n} \sum_{j=1}^{n} \mathbf{x}_j \tag{4.7}$$

Fig. 4.3 Between-clusters
distance based on global
centroid point

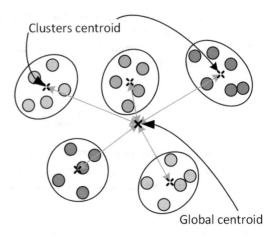

Clusters centroid

Global centroid

From scatter-matrix (4.6) it is possible to deduce the total within-cluster distance
as the trace of matrix S_W, tr$\{S_W\}$, which is equivalent to the cost function (4.4).
Using the second scatter-matrix (4.7), the distance between clusters is computed as
the tr$\{S_B\}$. The between-cluster distance measures the distance between each cluster
centroid and the global centroid of the data points; by maximizing this distance the
cluster separation is improved, Fig. 4.3.

The trace of both matrices can be used to define the KMS optimization problem
as:

$$\min \operatorname{tr}\{S_W\} \bigwedge \max \ \operatorname{tr}\{S_B\} \tag{4.8}$$

Since (4.8) defines a multi-objective optimization problem, in order to obtain just
one solution, it is common to convert the problem to a single-objective problem [10],
as in (4.9).

$$\min \operatorname{tr}\{S_W\}/\operatorname{tr}\{S_B\} \tag{4.9}$$

Two major factors affect the quality of KMS algorithm results. The first factor is
the parameter k, which defines the number of clusters. This parameter can assume
values from 2 to the total number of data points, where each data point is a cluster. Of
course, the extreme case where every data point is a cluster is not interesting, since
all data is considered dissimilar and no data reduction is achieved. The second factor
is how to initialize the clusters centroids. The initial location of cluster centroids
affects the convergence of the algorithm, which typically causes that two runs of
KMS reach different cluster results, because KMS, usually, gets trapped in local
minima. Since the two factors are common to all partitional clusters presented in this
work, both topics will be discussed after presenting all partitional cluster techniques.

The computational time complexity of the KMS algorithm is $O(nkdi)$ [11], where
n is the number of data points to cluster, k is the number of clusters, d is the space

dimensionality of the data, and i is the number of iterations performed until convergence, which in the case of Algorithm 4.1 has a maximum value of $R_{\max}$.

Algorithm 4.1 K-Means Algorithm

Input parameters:
    k – number of clusters;
    R_{max} – maximum number of iterations;
    ε – small number ($0 < \varepsilon < 1$).

1   Initialize cluster centroids c_i, $i=1\ldots k$
2    $r \leftarrow 0, J_{KMS}^{(r)} \leftarrow \infty$
3   Repeat
4   Compute the Euclidian distance of every data point x_j, with $j = 1\ldots n$. to every centroid c_i.
        $dist(i,j) = x_j - c_i^2; i = 1\ldots k, j = 1\ldots n$
5   Based on the computed distance $dist(i,j)$ allocate data point x_j to the closest cluster represented by c_i.
6   Update centroids positions by using (4.5)
7        $r \leftarrow r + 1$
8   Compute the cost function $J_{KMS}^{(r)}$ using (4.4)
9   Until $\left| J_{KMS}^{(r)} - J_{KMS}^{(r-1)} \right| < \varepsilon$ and $r \leq R_{max}$

4.2.2 K-Medoids Clustering Algorithm

The k-medoids (KMD) clustering algorithm [12] is very similar to the KMS algorithm. KMD, also, uses the same cost function, $J_{KMD} = J_{KMS}$, to be minimized, where J_{KMS} is defined in (4.4). The main difference is that KMS algorithm defines the cluster centroid as the representative element of the cluster, whereas the KMD algorithm adopts as cluster "center" the data point closest to the cluster centroid (4.10).

$$c_i = \arg \min_{x_j \in C_i} \left\| x_j - \sum_{p=1}^{n} 1_{C_i}(x_p) x_p / \sum_{p=1}^{n} 1_{C_i}(x_p) \right\| \tag{4.10}$$

The difference between the two clustering algorithms is clearly depicted in Fig. 4.4.

Due to the similarities between KMS and KMD, the only change in the KMD algorithm is in line 6 of Algorithm 4.2 where the new cluster "center" is selected. The time complexity per iteration of the KMD algorithm is $O(k(n-k)^2)$ [13], where k is the number of clusters and n is the number of data points.

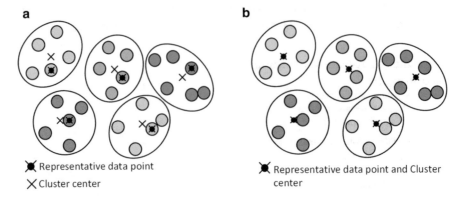

a

b

✖ Representative data point

✕ Cluster center

✖ Representative data point and Cluster center

Fig. 4.4 Difference between (**a**) K-medoid clustering algorithm and (**b**) K-means clustering algorithm

Algorithm 4.2 K-Medoid Algorithm

Input parameters:

    k – number of clusters;

    R_{max} – maximum number of iterations;

    ε – small number ($0 < \varepsilon < 1$).

1   Initialize cluster centroids c_i, $i=1\ldots k$

2    $r \leftarrow 0, J_{KMD}^{(r)} \leftarrow \infty$

3   Repeat

4   Compute the Euclidian distance of every data point $\mathbf{x}_j$, with $j = 1\ldots n$. to every centroid c_i.

        $dist(i,j) = \mathbf{x}_j - c_{i2}^2; i = 1\ldots k, j = 1\ldots n$

5   Based on the computed distance $dist(i,j)$allocate data point $\mathbf{x}_j$ to the closest cluster represented by c_i.

6   Update the c_i element by using (4.10)

7        $r \leftarrow r + 1$

8   Compute the cost function $J_{KMD}^{(r)} = J_{KMS}^{(r)}$ using (4.4)

9   Until $\left| J_{KMD}^{(r)} - J_{KMD}^{(r-1)} \right| < \varepsilon$ and $r \leq R_{max}$

4.2.3   Fuzzy c-Means Clustering Algorithm

The main difference of fuzzy c-means (FCM) clustering algorithm [14] to the already presented clustering algorithms is that FCM does not apply a hard or crisp clustering technique, which occurs when a data point is assigned to only one cluster. The FCM algorithm relaxes the definition of hard clustering; thus, a data point belongs to all clusters with a certain degree of membership. This characteristic is useful for problems where data points are close and not clearly well separated, resulting in a situation where data points may have high similarities to neighbor clusters. A comparison between FCM and KMS is presented in Fig. 4.5, where each

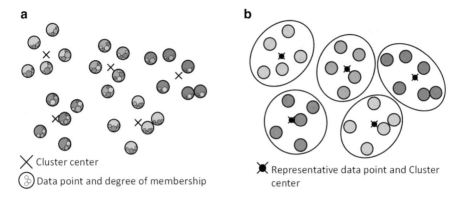

Fig. 4.5 (a) Fuzzy c-means vs. (b) K-means clustering algorithm

data point in the FCM algorithm has contributions from the other clusters based on the distance to that clusters. The degree of membership is depicted in Fig. 4.5a as inner circles, with color/gray tones different than the color/gray tone of the main cluster to which the data point belongs, i.e., the higher membership value.

Since FCM allows a data point to belong to more than one cluster, the previous formulation from KMS must suffer some minor changes. The FCM cost function is a fuzzy version of (4.4), where the distance from each data point to every cluster center is weighted:

$$J_{\text{FCM}} = \sum_{i=1}^{k}\sum_{j=1}^{n} \left(u_{ij}\right)^m \cdot \left\| \mathbf{x}_j - \mathbf{c}_i \right\|_2^2 \qquad (4.11)$$

where $u_{ij} \in [0, 1]$ represents the degree of membership of data point $\mathbf{x}_j$ to cluster i. The input parameter m defines the level of FCM fuzziness. Like the previous clustering techniques presented, this algorithm minimizes the within-cluster distances expressed by (4.11).

The degree of membership can be represented as matrix $U_{k \times n}$, where each element u_{ij} is calculated, [15], by (4.12):

$$u_{ij} = \frac{1}{\sum_{p=1}^{k} \left(\frac{\mathbf{x}_j - \mathbf{c}_i{}_2^2}{\mathbf{x}_j - \mathbf{c}_p{}_2^2}\right)^{1/(m-1)}}, \quad i = 1, \ldots, k; j = 1, \ldots, n$$

s.t.

$$u_{ij} \in [0, 1]$$

$$\sum_{i=1}^{k} u_{ij} = 1, \quad \forall j \in 1, \ldots, n \qquad (4.12)$$

$$0 < \sum_{j=1}^{n} u_{ij} < n, \quad \forall i \in 1, \ldots, k$$

From (4.12) results that the m fuzziness parameter must be greater than one ($m > 1$), a typical value is 2, and, on the limit when $m \to 1$, the membership value converges to 0 or 1, like in KMS or KMD. Considering (4.12), two exceptions must be added when computing the degree of membership:

$$
u_{ij} = \begin{cases} 1 & \text{if } \mathbf{x}_j - c_{i2}^2 = 0 \\ 0 & \text{if } \mathbf{x}_j - c_{p2}^2 = 0 \text{ and } p \neq i \end{cases}
\tag{4.13}
$$

The cluster centers adopted by FCM are computed as:

$$
\mathbf{c}_i = \frac{\sum\limits_{j=1}^{n} u_{ij}^m \cdot \mathbf{x}_j}{\sum\limits_{j=1}^{n} u_{ij}^m}, \quad i = 1, \ldots, k
\tag{4.14}
$$

Based on (4.11), the FCM within-cluster and between-cluster scatter matrix [16] are defined, respectively, by (4.15) and (4.16).

$$
S_{\text{FW}} = \sum_{i=1}^{k} \sum_{j=1}^{n} \left(u_{ij}\right)^m \left(\mathbf{x}_j - \mathbf{c}_i\right)\left(\mathbf{x}_j - \mathbf{c}_i\right)^T
\tag{4.15}
$$

$$
S_{\text{FB}} = \sum_{i=1}^{k} \sum_{j=1}^{n} \left(u_{ij}\right)^m \left(\mathbf{x}_j - \mathbf{C}\right)\left(\mathbf{x}_j - \mathbf{C}\right)^T \quad \text{where } \mathbf{C} = \frac{1}{n} \sum_{j=1}^{n} \mathbf{x}_j
\tag{4.16}
$$

The typical FCM [17] approach is to minimize the tr$\{S_{\text{FW}}\}$ by mutual update of (4.12) and (4.14). Several works, while minimizing the tr$\{S_{\text{FW}}\}$, also maximize the tr$\{S_{\text{FB}}\}$ to promote cluster separation [5, 16, 18].

Previously identified problems in the KMS and KMD algorithm also affect the FCM algorithm, such as the optimal number of clusters and initial cluster centers definition. Additionally, FCM requires defining the value for the fuzziness parameter m [19]. To understand the impact of the m parameter in the FCM algorithm, consider a particular iteration of the algorithm, data point $\mathbf{x}_1$ and that the number of clusters was set to three. Also, consider the squared Euclidian distances $D_{i1}^2 = \|\mathbf{x}_1 - \mathbf{c}_i\|_2^2$, presented in Fig. 4.6, with values detailed in Table 4.1.

The values presented in Table 4.1 allow computing the membership degree of $\mathbf{x}_1$ to each of the three cluster, assuming different values of m, Table 4.2.

From Table 4.2 it is possible to deduce the evolution of the degree of membership of $\mathbf{x}_1$ for an increasing value of parameter m. The degree of membership with respect to each cluster converges to the value 1/3, more generic for $\lim\limits_{m \to \infty} u_{ij} = 1/k$. This effect becomes clear in Fig. 4.7 where at each increment in parameter m the line becomes flatter and tending to a horizontal line with vale 1/3.

Fig. 4.6 Data point $\mathbf{x}_1$ distances to the 3 cluster centers

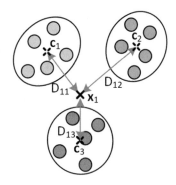

Table 4.1 Squared Euclidian distances between $\mathbf{x}_1$ and the cluster centers

D^2_{11}	D^2_{21}	D^2_{31}
1.5	2	1

Table 4.2 Degree of membership of $\mathbf{x}_1$ to each cluster for different fuzziness parameter values

m	u_{11}	u_{21}	u_{31}
2	0.307692	0.230769	0.461538
3	0.323544	0.280197	0.396259
5	0.329241	0.306393	0.364365
10	0.331716	0.321281	0.347003
20	0.332608	0.327610	0.339782
50	0.333061	0.331111	0.335828

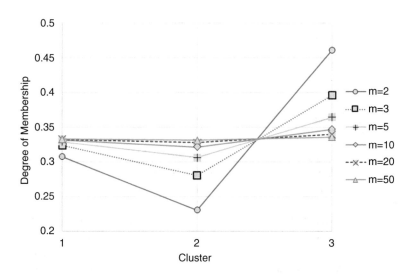

Fig. 4.7 Degree of membership for an increasing fuzziness parameter value

As the m parameter value increase the membership values tend to $1/k$, so data points start belonging equally to all clusters, and the clusters found become less significant [20]. In [21], the author suggested that the fuzziness parameter should be defined as $m \in [1.5, 4]$; also, tests for different m values showed that for noisy data the higher value of the interval presents more robust solutions. The adopted fuzziness parameter will always depend on the data, so several tests have to be performed before selecting the m value.

The FCM algorithm time complexity is $O(nk^2 di)$ [22], where n is the number of data points to cluster, k is the number of clusters, d is the space dimensionality of the data, and i is the number of iterations performed until convergence. Finally, to conclude this clustering algorithm description, the FCM algorithm is presented, Algorithm 4.3.

Algorithm 4.3 Fuzzy c-Means Algorithm

Input parameters:
    k – number of clusters;
    m – fuzziness parameter;
    R_{max}– maximum number of iterations;
    ε – small number ($0<\varepsilon<1$).

1   Initialize cluster centroids c_i, $i=1\ldots k$
2    $r \leftarrow 0$, $J^{(r)}_{FCM} \leftarrow \infty$
3   Repeat
4   Compute the degree of membership matrix by using (4.12).
5   Update the cluster centers c_i by using (4.14)
6        $r \leftarrow r + 1$
7   Compute the cost function $J^{(r)}_{KFC}$ using (4.11)
8   Until $\left| J^{(r)}_{FCM} - J^{(r-1)}_{FCM} \right| < \varepsilon$ and $r \leq R_{max}$

4.2.4 Partitional Clustering Parameters

The implementation and use of the previous cluster algorithms involve taking some initial decisions and the setting of several parameters, such as the number of clusters and how to initially select the cluster centers.

One of the simplest and most widely adopted techniques for selecting the number of clusters is known as the Elbow method [23]. The Elbow method can be easily explained graphically, where some cost function, like (4.4) or (4.11), is evaluated for an increasing number of clusters, and the "optimal" number of clusters corresponds to a point where an increase in the cluster number does not correspond to a large decrease in the cost function, as shown in Fig. 4.8. Another graphical method to define the number of clusters is the silhouette method. This method was presented by

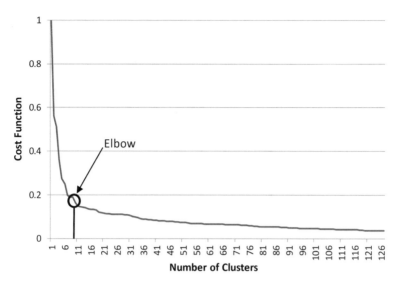

Fig. 4.8 Number of clusters using the Elbow method

Rousseeuw in [24]; the main idea is to identify if on average the data points are assigned to the correct clusters, so the silhouette value $s(i)$ acts as a validation cluster index. The number $s(i)$ is given by (4.17), where data point i is replaced by the nomenclature used so far for data points, $\mathbf{x}_j$.

$$s(\mathbf{x}_j) = \frac{b(\mathbf{x}_j) - a(\mathbf{x}_j)}{\max\left\{a(\mathbf{x}_j), b(\mathbf{x}_j)\right\}} \qquad (4.17)$$

where $a(\mathbf{x}_j)$ is the average dissimilarity of $\mathbf{x}_j$ to all the other data points in the same cluster, $b(\mathbf{x}_j)$ is the minimum dissimilarity of $\mathbf{x}_j$ to the data points in the rest of the clusters, and $\max\{a(\mathbf{x}_j), b(\mathbf{x}_j)\}$ returns the maximum value between $a(\mathbf{x}_j)$ and $b(\mathbf{x}_j)$. A typical dissimilarity measure is the Euclidian distance.

The silhouette index must be computed for all data points. A value close to one indicates that data point $\mathbf{x}_j$ was appropriately assigned to its cluster, because the value of $a(\mathbf{x}_j)$ is small indicating a good compactness of the cluster and $b(\mathbf{x}_j)$ has a high value signalling a suited cluster separation. Thus, the within-cluster distance has been minimized and the between-cluster distance has been maximized. The method must be applied for a different number of cluster and then select the number of clusters that on average provides results closer to one. To illustrate the silhouette method, consider 100 data points generated at random in two regions of a two-dimensional space, Fig. 4.9.

Performing two runs with the KMS clustering algorithm, using different number of clusters ($k = 2$ and $k = 5$), on the random data points in Fig. 4.9, reaches

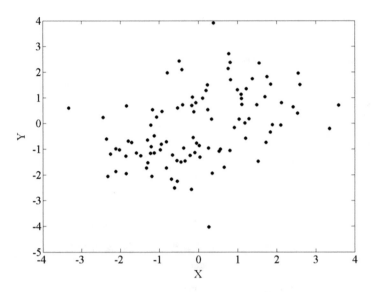

Fig. 4.9 One hundred random generated data points defining two clusters

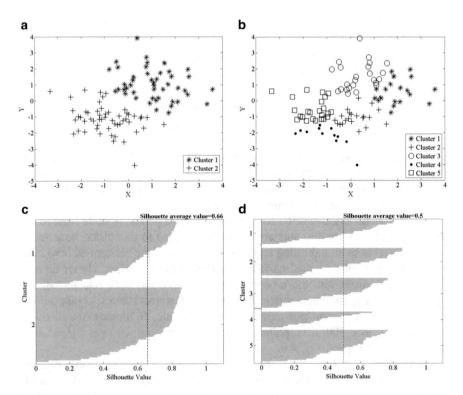

Fig. 4.10 (**a**) KMS algorithm with $k = 2$. (**b**) KMS algorithm with $k = 5$. (**c**) Silhouette graphic for two clusters KMS, with an average value of 0.66. (**d**) Silhouette graphic for five clusters KMS, with an average value of 0.5

Fig. 4.11 Random initial centers data points affect correct clusters identification

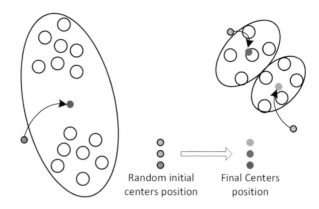

Random initial   Final Centers
centers position   position

the clusters formation in Fig. 4.10a for $k = 2$ run and Fig. 4.10b with $k = 5$. After executing the KMS algorithm, for both k settings, the silhouette index was computed. The obtained silhouette graphics are presented in Fig. 4.10c and d for $k = 2$ and $k = 5$, respectively. In the presented silhouette graphics, the line for the average silhouette index value is drawn. From the comparison of both average silhouette index values it is possible to conclude that two clusters explain better data points distribution, since for $k = 2$ the average index value is closer to one than for run with $k = 5$.

Many other approaches were developed to find the best number of clusters [25], but all of them require testing different number of clusters to discover the optimal cluster number, which may result in a computational expensive task.

The three clustering techniques already presented, KMS, KMD, and FCM, are very sensitive to initial cluster centers selection. The choice of different random initial center data points can affect both the convergence speed and cluster identification. In Fig. 4.11, a wrong cluster identification using random cluster centers initialization is shown.

The simplest approach to try to overcome the problem identified in Fig. 4.11 is to perform multiple runs of the clustering algorithm initializing randomly the cluster centers. Next the "best" run is selected by performing cluster validation. The cluster validation techniques allow to compare different clustering runs results by using cluster validity indexes and assess which of the runs is better in terms of found clusters [26]. Many cluster validity indexes adopt the concepts of within-cluster distance and/or between-cluster distances to estimate the quality of the results. The silhouette index is an example of a cluster validity index, using the within-cluster and between-cluster distances, which can be used to compare clustering runs. The Dunn index [26] estimates how compact and well separated are the clusters, which is defined by the ratio of the nearest cluster distance between C_q and C_r, and the maximum cluster C_p diameter (4.18):

$$\text{Dunn} = \frac{\min\limits_{1\le q\le k}\ \min\limits_{\substack{1\le r\le k \\ q\ne r}}\ \text{dist}(C_q, C_r)}{\max\limits_{1\le p\le k}\text{diam}(C_p)} \qquad (4.18)$$

where

$$\text{dist}(C_q, C_r) = \min_{\substack{\mathbf{x}\in C_q \\ \mathbf{y}\in C_r}} \|\mathbf{x}-\mathbf{y}\|_2 \qquad (4.19)$$

$$\text{diam}(C_p) = \max_{\mathbf{x},\mathbf{y}\in C_p}\|\mathbf{x}-\mathbf{y}\|_2 \qquad (4.20)$$

The distance between two clusters corresponds to the minimum distance between any data points in each of the clusters, in (4.19) the Euclidian distance was adopted but another dissimilarity measure can be used. The cluster diameter (4.20) is the maximum distance between any two points belonging to a cluster. A validity index adopted in the FCM algorithm is the partition coefficient [27], where the degree of membership is used to compute the index value:

$$\text{PC} = \frac{1}{n}\sum_{i=1}^{k}\sum_{j=1}^{n}u_{ij}^2 \qquad (4.21)$$

The partition coefficient quantifies the amount of overlap between clusters, the PC values are in the range $[1/k, 1]$, where for PC $= 1$ the solution corresponds to a crisp cluster partition. So, for higher values of PC better defined clusters are obtained. Whereas for values of PC close to $1/k$ indicates that the clusters are not well defined. Using this index is also possible to search for the optimal k parameter (cluster number), this is achieved by solving the optimization problem: $\max_{2\le c\le n-1}\text{PC}(k)$, where the partition coefficient is expressed in terms of a variable number of clusters.

Another validity index using the FCM degree of membership is the partition entropy index [28]:

$$\text{PE} = -\frac{1}{n}\sum_{i=1}^{k}\sum_{j=1}^{n}u_{ij}\log u_{ij} \qquad (4.22)$$

The partition entropy index, also, measures cluster overlapping. The range of values of this index is in the interval $[0, \log k]$. A lower cluster overlapping occurs for values close to 0, while for values at upper bound of the range indicates that clusters are not well defined. As in the partition coefficient this index can also be used to find the optimal k value, by minimizing the value of the index: $\min_{2\le c\le n-1}\text{PE}(k)$. These two cluster validity indexes, i.e., partition coefficient and partition entropy, are used later to validate a technique that reduces the number of clusters in the new yield estimation methodology. Many other cluster validity indexes were developed to assess the clustering process quality and, also, help finding the optimal number of clusters [29–31].

In order to avoid the random center initialization, a different number of techniques that look into the data points before defining the initial centers position start to appear. The linear assignment algorithm (LAA) [32] explores the concept of selecting the most dissimilar data points as cluster initial centers and has been used in FCM and KMS as initialization technique in several works [33, 34]. The LAA is an iterative technique that starts by finding the two most dissimilar representative data points of all the n data points (4.23); the dissimilarity measure is typically based on the Euclidian distance. The following iterations finds the more dissimilar data points to the already defined centers, until all the k initial centers are found (4.24).

$$\{\mathbf{r}_1, \mathbf{r}_2\} = \arg\max_{1 \leq i,j \leq n} \|\mathbf{x}_i - \mathbf{x}_j\|_2 \tag{4.23}$$

$$\mathbf{r}_l = \arg \max_{1 \leq i \leq n} \min\{\|\mathbf{x}_i - \mathbf{r}_1\|_2, .., \|\mathbf{x}_i - \mathbf{r}_{l-1}\|_2\}; \quad l = 3, \ldots, k \tag{4.24}$$

$$\mathbf{x}_i \neq \mathbf{r}_1, \ldots, \mathbf{x}_i \neq \mathbf{r}_{l-1}$$

The k-means++ [35] is a technique that also picks data points as initial centers for the KMS algorithm. This method initially selects a random data point as first cluster center. The rest of cluster centers are iteratively selected among the other data points, based on a probability proportional to the distance between each data point and the already selected centers. In Algorithm 4.4, the k-means++ algorithm is presented.

Algorithm 4.4 K-Means++ Algorithm

Input parameters:
        k – number of clusters;
        R_{max} – maximum number of iterations;
        ε – small number ($0 < \varepsilon < 1$).

1.   K-means++ cluster centers c_i, $i=1\ldots k$
1.a   Select at random one center c_1 from the data points;
1.b $l \leftarrow 1$
1.c   Repeat
1.d   Compute minimum distance for every data point to all centers already chosen.
            $D(\mathbf{x}_j) = \min dist(\mathbf{x}_j, c_i)$; $1 \leq j \leq n$; $1 \leq i \leq l$; $\mathbf{x}_i \neq c_1, \ldots, \mathbf{x}_i \neq c_l$
1.e $l \leftarrow l + 1$
1.f   Select $\mathbf{x}_j$ as center c_l with probability:
$$P(\mathbf{x}_j) = \frac{D(\mathbf{x}_j)^2}{\sum_{m=1}^{n} D(\mathbf{x}_m)^2}$$
1.g   Until $l < k$

2.   Proceed to the k-means clustering algorithm

Since the selection of the first center is at random and the following cluster centers selections has a probability associated, different runs of the k-means++ will achieve different initial centers results. A comparison between k-means++ and the LAA technique is depicted in Fig. 4.12.

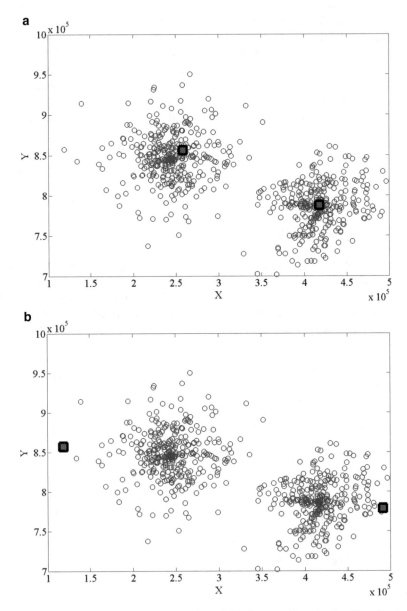

Fig. 4.12 Comparison between (**a**) k-means++ and (**b**) linear assignment algorithm for cluster centers initialization

In Fig. 4.12a two clusters example with 652 data points is presented. The initial cluster centers are identified as filled squares. The LAA approach since is based on the maximum Euclidian distance always selects the same distant pair of data points as initial centers. The k-means++ selects at random the first center, which due to the data distribution in the example is from one of the clusters. Since distant data points from the already select center have higher probability, it is likely that the second center is on the second cluster. In several runs performed the k-means++ initial centers were always data points from each of the clusters and close to the final clusters centroid than the initial cluster center of LAA, which results in a faster convergence for the k-means++ algorithm. Due to the similarity between KMS and FCM, the k-means++ initial center selection idea was incorporated into the FCM algorithm in [15]. Many other cluster initialization techniques adopted in KMS were converted to the FCM algorithm. A detailed comparison among different initialization techniques exported from KMS to FCM is presented in [36].

4.2.5   Hierarchical Clustering Algorithm

Hierarchical clustering has a different approach than partitional clustering techniques to cluster data. In hierarchical clustering, a set of nested clusters are organized as a hierarchical tree structure known as dendrograms, Fig. 4.13.

Hierarchical clustering has two different types of algorithms [37]. The agglomerative algorithm, which develops a bottom-up technique to define the cluster structure, starting by considering data points as individual clusters and at each iteration of the algorithm merges the most similar or closest pair of clusters. The second type of algorithm, known as divisive, has a top-down technique for defining the clusters. The hierarchical divisive clustering starts with one large cluster and at each step of the algorithm splits a cluster until only individual data points remain. An

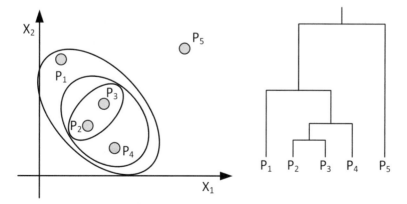

Fig. 4.13 Hierarchical clustering example

example of hierarchical divisive clustering is the bisecting k-means algorithm [38, 39]. The bisecting k-means algorithm, at each iteration, divides a larger cluster into two clusters using the KMS algorithm with $k = 2$. The criterion to select which cluster to divide may be the cluster with the largest variance or based on the number of data points in the cluster [40].

The hierarchical agglomerative clustering algorithm can be classified based on how the algorithm measures the distance between clusters, in order to select which clusters to merge. The most common distance measure techniques, according to [41], are:

1. Single-linkage, Fig. 4.14. This technique defines the distance measure between two clusters, C_q and C_r, as the minimum distance between a pair of data points, $\mathbf{x}$ and $\mathbf{y}$. In (4.25) the distance measure considered between data points is the Euclidian distance, but other metrics can be adopted, such as the Manhattan distance, Mahalanobis.

$$\text{dist}(C_q, C_r) = \min_{\substack{\mathbf{x} \in C_q \\ \mathbf{y} \in C_r}} \|\mathbf{x} - \mathbf{y}\|_2 \tag{4.25}$$

2. Complete-linkage, Fig. 4.15. The distance measure between clusters adopted by this technique is the maximum distance between a pair of data points:

$$\text{dist}(C_q, C_r) = \max_{\substack{\mathbf{x} \in C_q \\ \mathbf{y} \in C_r}} \|\mathbf{x} - \mathbf{y}\|_2 \tag{4.26}$$

Fig. 4.14 Single-linkage hierarchical agglomerative clustering

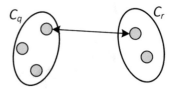

Fig. 4.15 Complete-linkage hierarchical agglomerative clustering

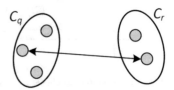

Fig. 4.16 Average-linkage hierarchical agglomerative clustering

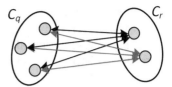

Fig. 4.17 Centroids distance hierarchical agglomerative clustering

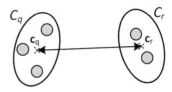

3. Average-linkage, Fig. 4.16. This technique considers the average distance between all pairs of data points from the two clusters:

$$\text{dist}(C_q, C_r) = \frac{1}{|C_q||C_r|} \sum_{\mathbf{x} \in C_q} \sum_{\mathbf{y} \in C_r} \|\mathbf{x} - \mathbf{y}\|_2 \qquad (4.27)$$

Another possible measure of distance between two clusters is considering the distance between the centroids of each cluster, Fig. 4.17 [42], given by (4.28), which is the measure adopted in this work for the agglomerative clustering algorithm.

$$\text{dist}(C_q, C_r) = \|\mathbf{c_q} - \mathbf{c_r}\|_2 \qquad (4.28)$$

where $\mathbf{c_q}$ and $\mathbf{c_r}$ are the centroids of clusters C_q and C_r, respectively.

Hierarchical clustering has two major advantages when compared to partitional clustering. First, it does not require a cluster number definition beforehand. Second, it only requires a matrix of distances between pairs of data points to perform the clustering task.

The time complexity of a hierarchical agglomerative clustering, where the number of scans in the matrix of distances for finding the most similar pairs of clusters is reduced at each iteration, is $O(n^2 \log n)$ [11], where n is the number of data points.

Algorithm 4.5 Hierarchical Agglomerative Clustering Algorithm

Input parameters:
      D – matrix with distance between pairs of data points.

1.   Repeat
2.   Find in D the most similar i,j clusters based on the distance measure adopted.
3.   Save distance between cluster i, j
4.   Merge clusters i and j.
5.   Delete in matrix D the rows and columns corresponding to cluster i and j.
6.   Compute the distance between the new merged cluster and the rest of clusters in D.
7.   Add the column and row with the distances computed in 5.
8.   Until D has only one element.

Algorithm 4.5 is capable of defining the hierarchical cluster dendrogram as in
Fig. 4.13. To achieve meaningful dimensionality data reduction from the hierarchical
clustering process is required to define a level where to cut the tree or a stopping level
criterion, to obtain a set of clusters that effectively represents the data. So, to perform
hierarchical clustering it is not required to specify the number of clusters at the start of
the clustering process, but after running the algorithm it is necessary to define an
approach which yields an appropriate number of clusters. The approach to decide
where to define the stopping level is typically based on the distance measure used to
decide which clusters to merge. To visualize the level where to cut the dendrogram, the
distance at each merge operation is saved and at the end of the clustering process it is
possible to plot this distance as the vertical axis of the dendrogram, in a manner that the
height of the node represents the similitude between merged clusters, Fig. 4.18.

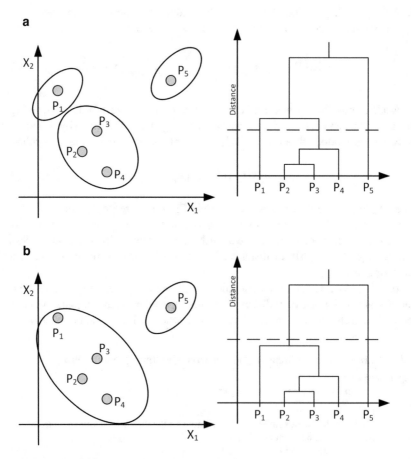

Fig. 4.18 Hierarchical clustering tree with different stopping levels affects the number of clusters
found. (**a**) Lower stopping level, 3 clusters were identified. (**b**) Higher stopping level, 2 clusters
were identified

Deciding at what level to stop the clustering process can be made by testing different levels and using cluster validation indexes [43]. Since all nested clusters in the hierarchy, from single data points to a lager clusters including all data, have been found by the algorithm, it is possible to perform several tests without repeating the clustering process. Other methods based on experimental observations to define the stopping level are the number of clusters, average dissimilarity within cluster, maximum distance between clusters, and clustering gain [44].

This work adopts as stopping criterion for the hierarchical agglomerative clustering algorithm a fixed distance level value between clusters that will be described in Sect. 4.3.5.

4.3   MC-Based Yield Estimation Using Clustering

Metaheuristic optimization strategies, such as GA, PSO, GSA, and NSGA-II, are iterative algorithms based on a large number of individuals, which represent potential solutions, that explores the search space towards the optimization goal. A general flowchart for metaheuristic evolutionary algorithms (EA) is presented in Fig. 4.19. In order to guide the exploration process, all potential solutions must be evaluated. At this stage of the algorithm, each potential solution objective value is calculated, and the problem constraints are assessed, which allows ranking the solutions accordingly. This rank allows the optimization algorithm to identify regions of the search space where the optimal solution(s) may reside.

Nowadays most automatic IC sizing and optimization tools adopt an electrical simulator in the evaluation stage to accurately estimate typical circuit performances

Fig. 4.19 Evolutionary algorithms flowchart

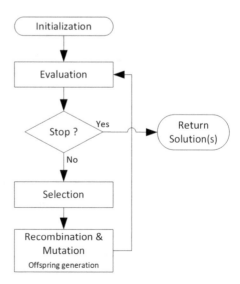

of each potential solution. Since one of the goals of this work is to implement an
MC-based yield estimation technique in a state-of-the-art simulation-based optimization IC sizing tool, the new methodology is to be included at the evaluation stage
to define the yield of each potential solution, without disturbing the rest of the flow.
To reduce the time impact of the additional simulations to estimate the yield of
potential solutions and since the implemented evaluation module already estimates
typical circuit performances, an ISE technique will be developed to avoid
performing MC simulations in potential solutions that do not satisfy the problem
constraints even under more favorable typical conditions. Whereas the solutions that
satisfy constraints under typical conditions are the input of the new yield estimation
methodology, according to the proposed flow presented in Fig. 4.1.

4.3.1 Infeasible Solution Elimination Module

The developed ISE module has two main goals. First, remove infeasible potential
solutions from being further simulated in the yield estimation module, and, second,
assign a yield value for those infeasible solutions in order to completely evaluate all of
them. Although it is possible assigning a zero value for the yield objective of all
infeasible solutions, the fact is that assigning the same yield value to all potential
solutions does not help with the convergence of the EA, because all potential solutions
will be equally bad. To overcome this problem each infeasible potential solution must
have a yield value assigned. Also, the yield value must be meaningful so that better
potential solutions have better yield values to help in the algorithm convergence. Based
on these ideas the new ISE module will use as data input the partial evaluated
population, output of the previous existing evaluation module. The ISE module then
classifies the potential solutions as *feasible* or *infeasible* based on the values already
estimated by the electrical simulator in typical conditions. This classification process is
easily performed by examining which potential solutions comply with the optimization
problem constraints, the feasible solutions, and which do not comply with the constraints, the infeasible potential solutions. Also, using the circuit typical performance
measures, estimated by the electrical simulator, was developed a metric that estimates
how far the infeasible potential solutions are from the feasibility region. Based on this
metric it was possible to compute a yield parameter value, which makes possible rank
all the population including infeasible potential solutions by assigning a meaningful
value to the yield parameter.

A circuit sizing optimization problem where the yield was added as one of the
objectives (4.29) and where only inequality constraints exists (4.30) is considered. If
equality constraints exist, they are easily converted into inequalities using (2.6).

$$\begin{cases} \min\ f_i(\mathbf{x}_d); \quad i = 1, \ldots, \#\text{Objectives} - 1 \\ \max\ \text{Yield}(\mathbf{x}_d) \end{cases} \tag{4.29}$$

s.t.

$$g_j(\mathbf{x_d}) \leq G_j; \quad j = 1, \ldots, \#\text{Inequalities} \tag{4.30}$$

In (4.30), G_j is a constant value that corresponds to constraint j boundary feasibility value, which allows classifying solutions as feasible or infeasible. In order to define a yield value for each infeasible potential solution, a boundary violation value is computed for each constraint, with its value being proportional to how much the solution overcomes the boundary value:

$$\text{BV}_j = \begin{cases} \dfrac{g_j(\mathbf{x_d}) - G_j}{G_j + \varepsilon}, & \text{if } g_j(\mathbf{x_d}) > G_j; \quad j = 1, \ldots, \text{Inequalities} \\ 0, & \text{otherwise} \end{cases} \tag{4.31}$$

Typically, G_j is different than zero and the $g_j(\mathbf{x_d})$ values are close to G_j, but to avoid division by zero errors a small term ε was added to the divisor. Using the boundary violation value from (4.31) is computed a "yield" parameter value for each potential solution as:

$$Y = -\frac{1}{\#\text{Inequalities}} \sum_{j=1}^{\#\text{Inequalities}} \text{BV}_j \tag{4.32}$$

From (4.32) results that potential solutions with $Y = 0$ are feasible solutions, while for negative yield parameter values the potential solutions are infeasible, with a value proportional to distance between the constraint measure value, $g_j(\mathbf{x_d})$, from the previous evaluation module and the boundary value in the constraint inequality, G_j.

The complete ISE module, including the previous evaluation module, is presented in Fig. 4.20. The EA send each population element to be evaluated by the existing

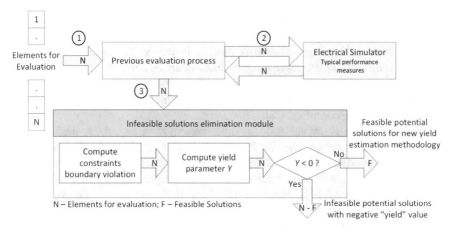

Fig. 4.20 Infeasible solutions elimination module

evaluation module, step 1. The previous evaluation module invokes the electrical simulator to estimate every required circuit performance measure (step 2). Having all elements partially evaluated, the N elements in the populations are used as input of the ISE module (step 3). Inside the ISE module the constraints boundary violation values are computed for each of the N elements of the population, and based on these values the yield parameter value is calculated for each potential solution according to (4.32). Using the yield parameter it is possible to classify as infeasible the elements with negative yield parameter value and as feasible otherwise. The ISE module outputs the F feasible elements to the new yield estimation methodology module, whereas the $N - F$ infeasible elements are already completely evaluated having a negative yield value proportional to how much the constraints are violated.

Feasible potential solutions at the output of the ISE module all have as yield a zero value, so the next step is to estimate the correct yield value for those elements. Next, a first approach using a k-means based methodology to estimate the yield is presented.

4.3.2 K-Means-Based Methodology for Yield Estimation

The new yield estimation methodology applies only to feasible potential solutions, since infeasible solutions were discarded by the ISE module, thus avoiding spending expensive simulation time on solutions that do not comply with the constraints under typical conditions. Considering the new evaluation module for yield estimation combined with the previous evaluation module and the ISE module, the complete evaluation process can be described as a two-step evaluation process. At the first step the elements of the population are evaluated in nominal conditions and the second step finishes the evaluation process considering variability conditions to estimate the yield of feasible solutions and assigning a negative yield value for infeasible potential solutions. As was presented in the proposed flow, Fig. 4.1, the reduction in the number of MC simulations to estimate the yield of potential solutions is achieved by performing those simulations just to a handful of potential solutions selected after performing a clustering process in the design variable space. The selected solutions are the representative individuals of each cluster, and the first clustering technique adopted and tested is the KMS clustering algorithm. A complete flow for the new two-phase evaluation process to be included at the optimization IC sizing tool is depicted in Fig. 4.21.

The *Clustering & RI selection* module, presented in Fig. 4.21, performs the clustering operation on the F feasible potential solutions and selects the C representative individuals, one from each of the clusters, which are simulated under variability conditions. After simulating the representative individuals, the *Yield Estimation* module analyzes the MC simulations and sets the estimated yield value by using (4.33) for the C-simulated potential solutions. Based on the yield values of the simulated solutions, the yield for the rest of the $F - C$ feasible potential solutions is estimated.

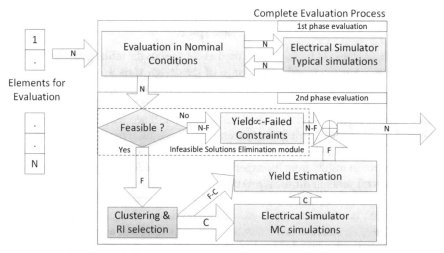

N – Elements for evaluation; F – Feasible Solutions; C – Number of Clusters

Fig. 4.21 New two-phase evaluation process for yield estimation

$$\widehat{Y} = \frac{C_{\#\mathrm{OkMC}}}{N_{\#\mathrm{MC}}} \tag{4.33}$$

where

$C_{\#\mathrm{OkMC}}$—Number of MC iterations that fulfill all optimization constraints;
$N_{\#\mathrm{MC}}$—Total number of MC iterations.

The new yield evaluation process adopts the KMS clustering algorithm [9], which has an easy implementation and performs quite well with large amounts of data. The main drawbacks for this clustering algorithm are the dependence of the initial cluster centers, the curse of dimensionality of the data, and the a priori knowledge of the numbers of clusters. In this work the k-means++ [35] cluster center initialization process was adopted. Also, the number of clusters was defined by the Elbow method.

The implemented Elbow method automatically defines the number of clusters based on several initial tests, before starting the optimization, where the representation error cost function (4.34) is evaluated for a growing number of clusters considering the EA population size.

$$\mathrm{Rep}_{\mathrm{Error}}(k) = \sum_{i=1}^{k} \sum_{\mathbf{x}_d \in C_i} \mathrm{dist}(\mathbf{c}_i, \mathbf{x}_d)^2 \tag{4.34}$$

where

k—Number of clusters;
$\mathrm{dist}(\cdot)$—Euclidian distance between two points;
C_i—Cluster i;

c_i—Cluster i center;
$\mathbf{x}_d$—Data point vector from design variable space.

To compute the cost function were used as data points the initial population of the EA, randomly created based on the range of the design variables defined in the IC sizing and optimization problem. Then, the cost function computes the representation error using the initial population evaluated for a growing number of clusters from 2 to the number of elements of the population. Since the EA population size adopted to solve our IC sizing and optimization problems, typically, assumes values from 64 to 512 individuals, the clustering process is quite fast. The presented population range results from the experience of using the IC sizing and optimization tool for several years in different real optimization problems.

Based on the computed representation error, the variation in the representation error between consecutive runs is examined and the number of clusters is set to a value k, where an increase in the number of clusters results in a representation error variation below some threshold value, (4.35).

$$\Delta \mathrm{Rep}_{\mathrm{Error}} = |\mathrm{Rep}_{\mathrm{Error}}(k) - \mathrm{Rep}_{\mathrm{Error}}(k+1)| \leq \varepsilon \tag{4.35}$$

The obtained number of clusters defined by the representation error variation will be used and maintained during all the IC sizing and optimization processes.

The Euclidian distance adopted in the representation error function revealed a problem, which is shared by the clustering process since also adopts the same metric distance to cluster data. The problem is caused by the different magnitude of the several design variables considered for the IC sizing and optimization problem, resulting that higher magnitude variables dominate the Euclidian distance value. A typical example of this problem is when are considered variables such as voltages with a magnitude of units of volts and transistors gate widths with magnitude of tens of nanometers to compute the Euclidian distance. So, to solve this problem all design variables vectors components are scaled by the Min–Max normalization or data transformation technique:

$$x'_{d,i} = \frac{x_{d,i} - X_{d,i}^{(\min)}}{X_{d,i}^{(\max)} - X_{d,i}^{(\min)}} \tag{4.36}$$

where
$x_{d,\,i}$—Component of a design variable vector $\mathbf{x}_d = [x_{d,\,1} \ldots x_{d,\,n}]^T \in \mathbb{R}^n$ to normalize;
$X_{d,i}^{(\min)}$—Minimum value of all $x_{d,\,i}$ in the population;
$X_{d,i}^{(\max)}$—Maximum value of all $x_{d,\,i}$ in the population.

After performing the data transformation (4.36) all the components of the design variable vector are in the range [0, 1]. This scaling operation also helps cluster identification since KMS is based on the Euclidian distance assuming that clusters are hyper-spherical or globular, and with similar sizes [8, 45, 46].

The typical implementation of KMS selects the centroid of the cluster as the representative individual of each cluster. Creating a potential solution as representative individual of the cluster based on the centroid components, which most likely was not evaluated at the first phase of the evaluation process, may reveal a problem since this virtual solution is probably infeasible. To overcome this problem, a feasible potential solution from each cluster must be selected. The natural approach was, then, choosing as the cluster representative individual the closest solution to each cluster centroid. This idea is similar to perform the KMD clustering algorithm, but since the KMS algorithm was already implemented, a hybrid solution was implemented. The developed clustering solution performs the typical KMS algorithm by minimizing the cost function (4.4) and after defining all the clusters, a single iteration where the cluster centers are updated according to (4.10) is executed. Thus, the representative individual selected from each cluster is the feasible potential solution closer to the cluster centroid. The next step, after selecting the representative individuals, is performing MC simulations using the electrical simulator to accurately estimate the yield of those individuals or potential solutions. Having the estimated yield value of the simulated potential solutions, the *Yield Estimation* module has to assign a yield value for the rest of feasible potential solutions to conclude the evaluation process of the rest of feasible potential solutions in the population. Considering that all potential solutions in a cluster are similar, i.e., close to each other in the design variable space, and since the selected representative individual minimizes the representation error, results from the crisp degree of belonging of each individual to its cluster that all individuals or potential solutions in a cluster have the same yield value of the simulated representative individual.

The initial tests for multi-objective IC sizing and optimization using the described methodology revealed problems related to the selection of cluster representative individuals. The approach of choosing the closest potential solution to each cluster centroid as the cluster representative individual creates a problem, because the selected potential solution may have a higher yield value than some of the other potential solutions in the cluster, which results in the promotion of the yield value at the *Yield Estimation* module causing the appearance of a false better POF, Fig. 4.22. The example of Fig. 4.22 shows the false POF effect based on real data, and at early stages of the optimization process. The presented false POF reveals that, since all potential solutions have the same yield value of the tested cluster representative individual, some potential solutions have their yield value increased, creating a new POF with values better than the real or reference POF. This effect is clear for points such as solution 5 and 10 whose real yield value is 89 and 83, respectively, and after assigning the yield value of the cluster representative individuals have their yield increased.

The false POF effect is also observed at later stages of the IC sizing and optimization process. In Fig. 4.23 a different example shows the same problem at the end of the optimization process, where the stars markers represent the POF obtained by performing MC simulation in all individuals of the population, while the cross markers define the POF of the same population when applying the MC simulation to the closest to cluster centroid solution for a 10 clusters KMS run.

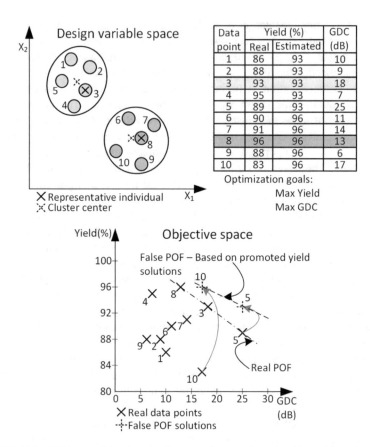

Fig. 4.22 False POF, caused by the crisp k-means degree of membership and by selecting the closest potential solutions to cluster center which promotes solutions to higher values than its real yield value

4.3.3 Solving the False POF Problem

The selection of potential solutions with higher yield values than the rest of potential solutions in a cluster as the cluster representative individuals undermines the optimization processes. By overestimating the yield of potential solutions, individuals in the EA population have their fitness increased resulting in the false POF. The false POF problem must be addressed, since solutions may reveal poor reliability on later stages of the IC design flow, forcing redesign iterations.

In order to correct the false POF problem, a new approach to select the cluster representative individual was developed. This new approach selects the best individual at each cluster in terms of the objective(s) function value(s). The objectives functions considered are all the objectives defined in the optimization problem except the yield, since at this stage of the evaluation algorithm the yield was not

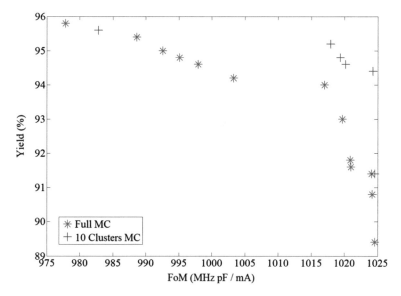

Fig. 4.23 False POF effect at the end of the optimization process, caused by selecting the element closest to the cluster center as the cluster representative individual, trade-off Yield (%) vs. Figure-of-merit (MHz pf/mA)

estimated. The definition of the best individual or potential solution at each cluster in terms of the objective function is easily defined for optimization problems with two objectives, being one of them the yield. At this bi-objective type of problems, the solutions are defined in a one-dimensional objective space, considering only the non-yield objective, and the best potential solution is selected based on the scalar value of the objective function computed at the first phase of the evaluation process. For more than two-objective optimization problems, a POF is defined based on the values of the objective functions' values computed at the first phase of the evaluation process. Since one potential solution in the POF cannot be considered better than other potential solution in the front, a best trade-off point is created based on the best values at each objective. Next, the potential solution closest to this ideal point is selected as the cluster representative individual, as illustrated on Fig. 4.24.

Applying the approach of selecting as cluster representative individual the best objective individual per cluster at the example in Fig. 4.22 showed that the new POF only has one solution, data point 5, since all the other potential solutions are projected into the line of the cluster representative yield value and all potential solutions become dominated by potential solution 5, Fig. 4.25.

This new approach of selecting the cluster representative individual, where the potential solutions considered are the one with the best objective(s) value(s) at each cluster, was possible since the circuit sizing and optimization problem usually involves improving several objectives different than the yield, and all those objectives are evaluated at the first step of the evaluation stage. The approach of selecting

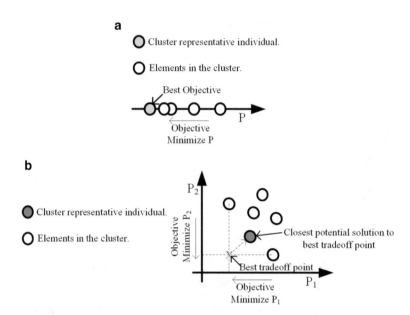

Fig. 4.24 Selection of the cluster representative individual based on the objective(s) value(s). (**a**) One-dimensional objective problem. (**b**) Two-dimensional objective problem

the potential solution with the best objective value per cluster presents the best choice since the algorithm will perform MC simulations on those potential solutions, which are the most likely to become members of the POF, as they are the best potential solutions regarding the problem objectives. Additionally, this option represents a worst-case cluster yield estimation scenario, because the best elements are expected to be closer to the limits or boundaries of the optimization problem feasibility region, and a slight parameters perturbation, like the ones caused by the MC iterations, may led to a higher number of failed iterations, thus lowering the yield.

In order to clarify the proposed modified KMS algorithm, an example using a test circuit from the results chapter is presented in Fig. 4.26. In this example, for simplicity, only two clusters and two variables were considered, the *W* and *L* of transistor M0. The objective value is the Figure-of-Merit (FoM) of the circuit, which is measured with the electrical simulator.

The KMS algorithm receives as parameters the solutions in the variable space and the corresponding objective function values. The KMS algorithm, then, performs the regular clustering operation, where it tries to minimize the Euclidian distance between the solutions in each cluster and the cluster centroid. After the clustering operation, the defined clusters are mapped into the objective space and the points with the maximum objective value, FoM, are identified on each cluster. The final step identifies back the points with maximum FoM into their respective points on the variable space, which are returned as the cluster representative individuals.

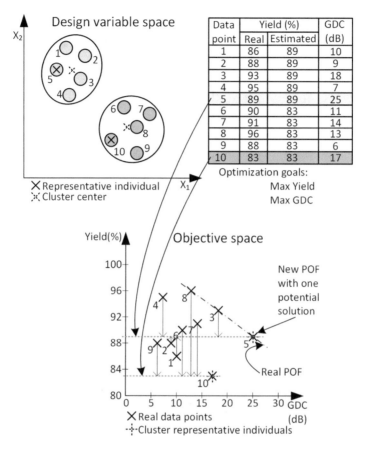

The following table appears within the figure:

Data	Yield (%)		GDC
point	Real	Estimated	(dB)
1	86	89	10
2	88	89	9
3	93	89	18
4	95	89	7
5	89	89	25
6	90	83	11
7	91	83	14
8	96	83	13
9	88	83	6
10	83	83	17

Fig. 4.25 Cluster representative individual selection based on the best objective(s) values

4.3.4   Projection of Potential Solutions into the Cluster Representative Individual Yield Line

Based on some initial tests for yield-aware optimization, using the methodology described in Sect. 4.3.2, several minor problems were identified. Although those problems were identified when using the KMS-based approach, they are not only related to the KMS algorithm but exists in all the clustering algorithms tested in the IC sizing multi-objective yield-aware optimization methodology presented in this work. This section discusses those problems, presents solutions, and tries to further reduce the number of MC simulations needed to estimate the yield of potential solutions inside the optimization loop.

Since one of the problems is related to the cluster representative individual selection, several other techniques to select the cluster representative individual are presented in this section. At this stage of the work, the preferred approach to select

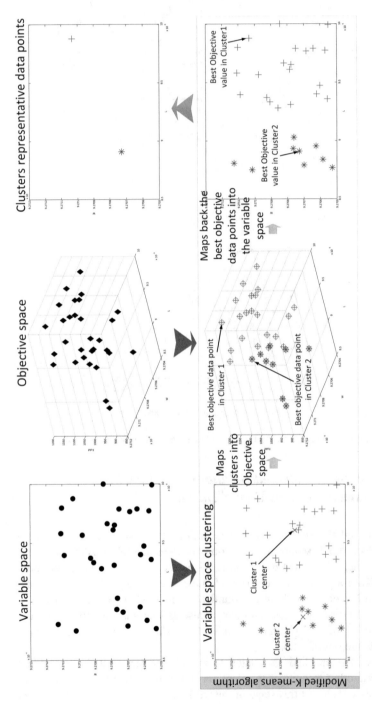

Fig. 4.26 K-means clustering algorithm for best objective value representative potential solution example

the cluster representative individual is identifying the potential solution with best objective(s) value(s) per cluster, computed at the first evaluation phase. Nevertheless, this preferred approach solves the problem of the false better POF, from Fig. 4.25 it is possible to observe another type of problem. In Fig. 4.25 is presented the POF when using the real yield values of the potential solutions; this POF has three potential solutions, data points 3, 5, and 8. But when the new approach for cluster representative individual selection is adopted, the POF only has one potential solution, data point 5, because all other potential solutions with higher yield values in the POF are discarded, since data point 5 dominates the rest of potential solutions.

The effect of discarding higher yield values is clearly shown in Fig. 4.27, where the final POF obtained by the new cluster representative individual approach is compared with the reference POF where all the solutions were submitted to MC analysis.

The POF in Fig. 4.27 reveals that at the higher yield region the two curves are moving apart, whereas for the lower yield region both POF solutions are similar. The problem is caused by the projection of all the cluster potential solutions into the yield line value of the cluster representative individual, since all potential solutions in a cluster will have the same yield value of the cluster representative individual. This effect underestimates the yield of several potential solutions in the cluster, as was earlier discussed, since cluster representative individuals with best objective(s) value (s) are typically closer to the boundary of the feasibility region. The projection of potential solutions into the cluster representative individual yield line effect is shown in Fig. 4.28.

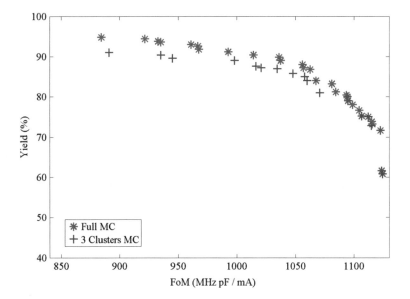

Fig. 4.27 Comparison between a 3-cluster yield optimization POF and the reference POF where all the elements were submitted to MC, trade-off Yield (%) vs. Figure-of-merit (MHz pf/mA)

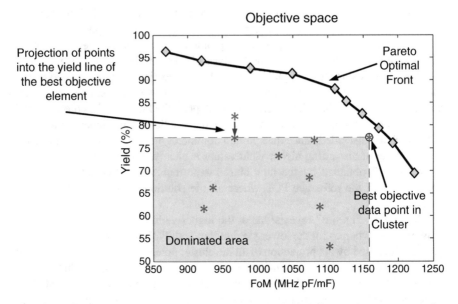

Fig. 4.28 Effect caused by the selection of the best objective solution as cluster representative individual

In Fig. 4.28 the stars represent a cluster of solutions with real yield values and the line with diamonds defines the current POF. This example shows that when selecting the best objective potential solution as representative individual, in this case the solution with best FoM, the rest of potential solutions of the cluster will be discarded by the multi-objective optimization algorithm (NSGA-II algorithm) since they are all dominated, even points with a better yield than the selected representative element. This fact explains the distance between the reference and the obtained POF when applying the new MC-based yield optimization process. One solution for this problem is to increase the number of clusters, as it is shown in Fig. 4.29, but a large number of clusters degrade the optimization speed performance, since more potential solutions must be submitted to MC simulations.

In order to solve this problem, two approaches were implemented and studied. The first approach addresses the cluster representative individual selection, where two new methodologies are proposed and detailed in the next Sect. 4.3.4.1. The second approach deals with the number of clusters used by the clustering algorithm and is presented in Sect. 4.3.4.2.

4.3.4.1   Cluster Representative Individual Selection

The first new method to select the cluster representative individual is based on the design centering technique discussed in Chap. 3. This technique finds higher yield potential solutions supported by the fact that solutions located well inside the feasibility region are more robust to the variability effects than solutions close to the boundaries, Fig. 4.30.

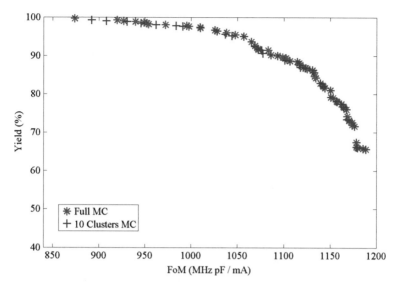

Fig. 4.29 Comparison between a 10-cluster yield optimization POF and the reference POF where all potential solutions were submitted to MC, trade-off Yield (%) vs. Figure-of-merit (MHz pf/mA)

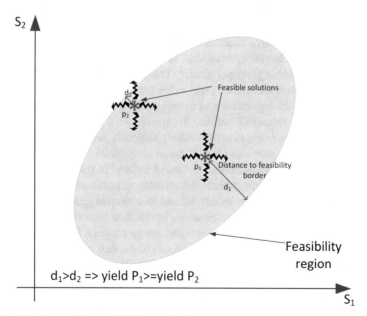

Fig. 4.30 Solutions inside the feasibility region have higher yield values than solutions near the boundary

Considering an IC sizing and optimization problem as in (4.29) and (4.30) was created a measure that computes the distance between the solution and the boundaries of the feasibility region defined by the problem constraints conditions. This measure allows to implement the new cluster representative individual selection methodology based on the Euclidian distance (4.37).

$$D = \sqrt{\sum_{j=1...\#Constraints} \left(g_j(\mathbf{x_d}) - G_j\right)^2} \qquad (4.37)$$

where

G_j—Limit value define by the j constraint of the optimization problem;
$g_j(\mathbf{x_d})$—Value estimated by the electrical simulator that corresponds to constraint j of solution $\mathbf{x_d}$.

Since some constraint conditions may have a greater impact than others on (4.37) distance, each measured constraint and limit value were scaled using the Min–Max technique (4.36). Therefore, the cluster representative individual is the solution per cluster that has the larger distance, i.e., higher D value. So, the higher distance to boundary method for selecting the cluster representative individual will use this new parameter D, like the objectives values were used in the previous discussed method, Fig. 4.26, where the best objective(s) solutions were selected as the representative individual. Using the KMS-based methodology results the flow presented in Fig. 4.31.

The cluster representative individual selection method based on the distance to the feasibility region boundary was able to identify in each cluster the potential solutions with the highest yield per cluster. The problem with this method was that potential solutions with the higher yield value typically present more conservative values for the rest of the objectives and as was previously discussed this may lead to the appearance of a false POF effect. To overcome this problem an alternative method for cluster representative individual selection was developed. This new approach combines the previous idea (distance to feasibility region boundary) with the best objective solution selection presented before. In this new method, KMS algorithm receives the optimization objective(s) value(s) and the new distance measure given by (4.37). Those values defined an objective space with an additional dimension, defined by the distance measure. In this new objective space, combining the optimization problem objectives and distance measure D, it is possible to define an ideal point with coordinates that corresponds to the best values per objective and larger distance measure, as illustrated in Fig. 4.32. Therefore, the cluster representative selected is the element closest to the ideal point.

These two new cluster representative individual selection methodologies were evaluated in the results chapter and compared with the reference selection methodology where the best objective(s) individuals per cluster are considered as the cluster representative individuals. The tests conclude that selecting the cluster representative individual using the distance to the feasibility boundaries results in the appearance of

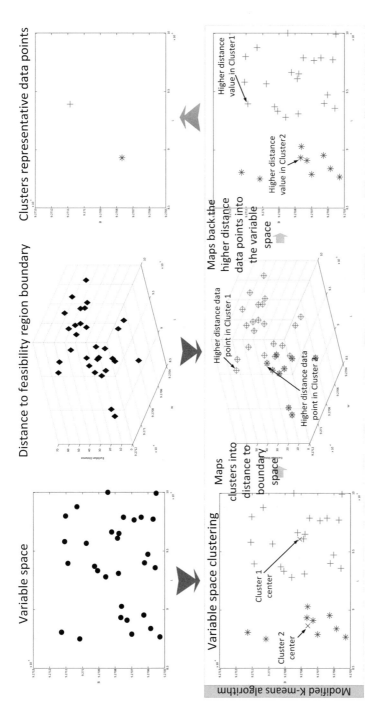

Fig. 4.31 Higher distance to feasibility region boundary method for cluster representative individual selection

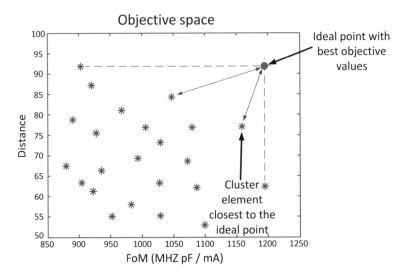

Fig. 4.32 Cluster representative individual selection combining distance to feasibility region boundaries and optimization problem objective(s)

a false POF problem. By adding the optimization problem objectives into the distance approach, the underestimating yield effect is marginally reduced, thus not justifying the adoption of this new cluster representative individual selection because of the additional computational time.

4.3.4.2   K-Means Variable Cluster Number Selection

The presented yield estimation methodology based on clustering adopts a fixed number of clusters defined by the Elbow method. As was previously discussed, if the number of clusters selected to perform the clustering process is too small, then the problem related to the projection of all potential solution in a cluster into the yield line value of the cluster representative individual may appear. The solution for this problem is to increase the number of clusters, which results in an increase of the number of potential solutions that must be subject to MC analysis. Therefore, this approach results in an increase of the execution time required to estimate potential solutions yield values.

The implemented Elbow method uses the random initial population of the GA-based optimization algorithm to define the number of clusters. Using just one sample of the population to run the Elbow method may result in a small number of clusters. Also, the number of clusters defined at the beginning of the optimization process may not be the most suitable cluster number for later stages of the optimization process. In order to overcome the difficulty of selecting the correct number of clusters were adopted two variable cluster number approaches. In both approaches the number of clusters is defined at each generation of the optimization process, which is related to the current generation number and the maximum defined generations. These two variable cluster numbers methods were based on the same exponential function, in the first case an exponential cluster number decay (4.38) was used and in the second an exponential growth function (4.39) was adopted.

$$C_d(i) = \left\lceil C_{max} e^{-\alpha \frac{i-f}{G-f}} \right\rceil; \quad f \le i \le G \tag{4.38}$$

$$C_g(i) = C_{max} + 2 - \left\lceil C_{max} e^{-\alpha \frac{i-f}{G-f}} \right\rceil; \quad f \le i \le G \tag{4.39}$$

where

C_{max}—maximum number of clusters;
α—defines the rate of decay or growth;
i—current generation number;
G—maximum number of generations;
f—generation number where the first feasible solution appears;
$\lceil x \rceil$—rounds upward the argument, returning the smallest integer value that is not less than x.

Both functions (4.38) and (4.39) require the definition of two parameters, which are the maximum cluster number parameter and the rate of decay or growth. To define the C_{max} parameter, the Elbow method is executed several times, at the beginning of the optimization process and at each run a different random initial population is generated. Then, the highest number of clusters among the different Elbow runs is selected as the C_{max} parameter. Whereas the α parameter value was defined as:

$$\alpha = \ln\left(\frac{C_{max}}{2}\right) + 0.5 \tag{4.40}$$

where $\ln(\cdot)$ refers to the natural logarithm function.

The α parameter in (4.40) has two terms, the logarithmic term results from solving a relaxed version of (4.38), where the upward round function was removed, in order to ensure that when $i = G$ the function $C_d(i)$ returns two clusters, while the second constant term was added to ensure that the exponential decay function reaches the two clusters value before the optimization process ends. Since the growth exponential function (4.39) is based on the same exponential function as (4.38), the same α parameter (4.40) was adopted. By selecting this α parameter in (4.39) was ensured that when the optimization process is reaching the end, the number of clusters return by $C_g(i)$ is the maximum cluster value, the C_{max} parameter. A graphical representation of both functions is depicted in Fig. 4.33.

The graphic examples of Fig. 4.33 assume that the optimization process will run until the stopping criteria of 300 generations is reached ($G = 300$) and that in the initial generation a feasible solution already exists ($f = 0$). The rest of parameters are

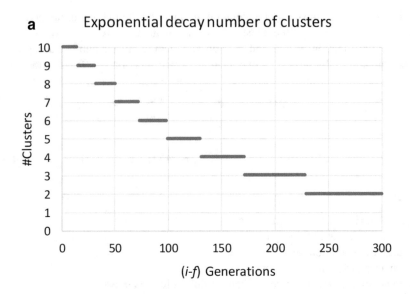

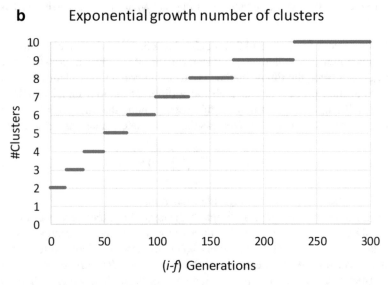

Fig. 4.33 Exponential (**a**) decay and (**b**) growth functions with $C_{max} = 10, f = 0, G = 300$

the $C_{max} = 10$ due to the different Elbow method runs and the α parameter that was set according to (4.40).

The adoption of one of the variable number of clusters approaches in the IC sizing and optimization process is capable of providing additional savings in the total number of MC simulations when compared to the typical reference approach, where a fixed number of clusters was defined for the overall optimization process.

Fig. 4.34 Number of MC
simulations relation between
the fixed cluster number
approach (full area of the
graphic) and variable
clusters number (dark gray
area of the graphic)
considering $C_{max} = 10$

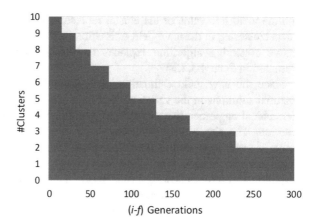

Considering the example shown in Fig. 4.33a it is possible to estimate the number of
MC simulations saved by adopting the variable cluster number approach. The
adoption of a fixed cluster number using the Fig. 4.33 example parameters requires
a total number of 3000 (10 clusters × 300 generations) potential solutions to be
simulated to estimate the yield, whereas for the variable cluster number the total
number of potential solutions simulated is 1394; hence the adoption variable cluster
number results in a saving of 53% of the MC simulations.

The relation of potential solutions simulated by each approach can be observed
graphically in Fig. 4.34. The total number of simulated potential solutions, when the
variable cluster number is adopted, is represented by the area identified in dark gray,
while the total number of potential solutions for the fixed approach corresponds to
the total area of the graphic.

The variable cluster number approaches were tested using the KMS clustering
algorithm. The tests show that the adoption of these approaches results in additional
savings of the total number of MC simulations when compared to a fixed number of
clusters during the overall optimization process. However, the reduction in the
number of potential solutions simulated increases the yield estimation error.

4.3.5   Hierarchical Agglomerative Clustering for MC-Based Yield Estimation

Partitional clustering algorithms, such as KMS and KMD, require the previous
setting of the number of clusters to perform the clustering task. Hierarchical clustering algorithms delay the definition of the number of clusters to the last step of the
algorithm, making possible deciding a posteriori the most suitable number of
clusters for the problem without repeating the clustering task. The possibility of
tuning the number of clusters at each generation of the optimization algorithm can
minimize the problem related to the projection of all potential solution in a cluster

into the yield line value of the cluster representative individual. For testing hierarchical clustering in the new yield estimation methodology, the hierarchical agglomerative clustering algorithm replaces the KMS algorithm in the methodology presented in Sect. 4.3.2.

Since the new yield estimation methodology is based on the fact that close potential solutions in the variable space have similar yield values, the implemented hierarchical agglomerative clustering algorithm adopts the single-linkage measure criterion. The single-linkage measure uses the minimum distance between data points as criterion to define clusters, Sect. 4.2.5. The distance measure used by the single-linkage criterion is registered at each step of the hierarchical agglomerative clustering algorithm, allowing to define the dendrogram and, also, create a rule of similarity to define the final number of clusters to consider at each generation of the optimization algorithm. The implemented hierarchical agglomerative clustering single-linkage measure is based on the Euclidian distance between potential solutions in the variable space. Since all design variables vectors components are scaled to the interval $[0, 1]$ using (4.36), the Euclidian distance is in the interval $[0, \sqrt{n}]$, where n is the number of design variables.

Tests using the AIDA-C sizing and optimization tool, with different test circuits, revealed that feasible solutions on later stages of the optimization process, i.e., the exploitation phase, have the same values on most of the design variables. To corroborate this situation, consider the circuit in Fig. 4.35 and two different sized solutions, which after optimizing the Figure-of-Merit (FoM) and the gain DC (GDC)

Fig. 4.35 Schematic of the tested single-stage amplifier with enhanced DC gain

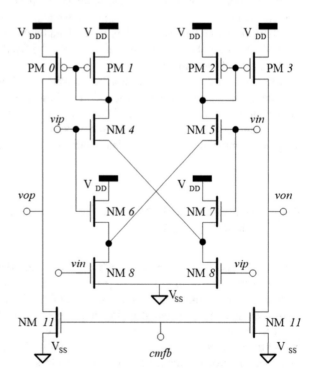

Table 4.3 Two sized solutions for the single-stage amplifier with enhanced DC gain circuit, Fig. 4.35

Solution 1

Design variables		Optimization Objectives
l0 = 670 nm	w0 = 56.0 µm	FoM = 980 MHz pF / mA
l1 = 300 nm	w1 = 5.2 µm	GDC = 53.23 dB
l10 = 940 nm	w10 = 3.8 µm	
l4 = 760 nm	w4 = 27.0 µm	
l6 = 820 nm	w6 = 88.6 µm	
l8 = 630 nm	w8 = 1.0 µm	
nf0 = 3	nf4 = 1	
nf1 = 3	nf6 = 5	
nf10 = 1	nf8 = 3	

Solution 2

Design variables		Optimization Objectives
l0 = 600 nm	w0 = 70.3 µm	FoM = 1015 MHz pF / mA
l1 = 300 nm	w1 = 5.6 µm	GDC = 52.32 dB
l10 = 940 nm	w10 = 5.3 µm	
l4 = 800 nm	w4 = 26.8 µm	
l6 = 820 nm	w6 = 94.1 µm	
l8 = 630 nm	w8 = 1.0 µm	
nf0 = 5	nf4 = 3	
nf1 = 1	nf6 = 5	
nf10 = 1	nf8 = 3	

The nf_ design variables correspond to the transistor number of fingers, w_ is the transistor channel width, and l_ is the transistor length channel

attained the presented design variables values. In Table 4.3, both solutions are detailed, and it is possible to verify that from the 18 design variables, eight of them have the same value, identified by the gray cells.

Based on the tests observations and considering that potential solutions with identical design variables values except in one of them are highly similar, the Euclidian distance dendrogram stopping level to define the number of clusters was initially set to 1.

The selection of a fixed distance measure results in a variable number of clusters. Since at the beginning of the sizing and optimization algorithm potential solutions are spread over the search space, the number of clusters is high because the solutions are very dissimilar. In later stages of the optimization algorithm, the GA population starts to converge to the final solution(s), resulting in a smaller cluster number. This effect is presented in Fig. 4.36 where the single-linkage Euclidian distance line level

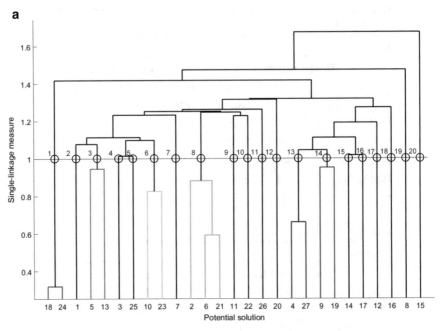

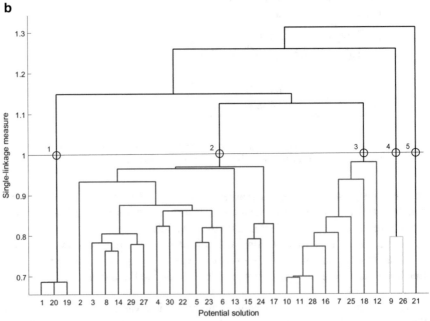

Fig. 4.36 Dendrogram at different stages of the sizing and optimization process. (**a**) Beginning of the optimization process, dissimilar solutions results in 20 clusters. (**b**) End of the optimization process, similar solutions results in 5 clusters

of 1 is presented. The intersection between the dendrogram lines and the defined Euclidian line level, identified by a circle, represents a cluster. Comparing Fig. 4.36a, b two points stand out. First, the number of clusters at early stages of the optimization process is four times greater than the number of clusters at the end of the optimization process. Second, the number of clusters with just one potential solution indicating that solutions are less similar at the beginning of the optimization processes than at the end.

The new yield estimation methodology using hierarchical agglomerative clustering, like in the previous KMS-based methodology, selects the cluster representative individual as the potential solution in each cluster with the best objective(s) value (s) to avoid false POF problems.

The adoption of the hierarchical agglomerative clustering algorithm, having a variable number of clusters according to the similarity of potential solutions, results in a better accuracy of the yield estimation than in the KMS-based methodology. The main reason for the accuracy improvement is related to the fact that on average this methodology uses a higher number of clusters per generation than the KMS-based methodology. The higher number of clusters represents additional expensive MC simulations. Since most of the simulated potential solutions become dominated by other simulated potential solutions, other Euclidian stopping levels were tested. The different tests, using a Euclidian distance value higher than one, revealed worst accuracy than the KMS-based methodology for the same number clusters.

4.3.6 Fuzzy c-Means Application for Accurate and Efficient Analog IC Yield Optimization

The methodologies for yield estimation presented in Sects. 4.3.2 and 4.3.5 adopt clustering techniques with a crisp degree of membership. From the crisp degree of membership results that all elements of a cluster have the same yield value of the representative cluster element, which can affect the accuracy of the yield estimation for the non-MC-simulated potential solutions. As was shown in the hierarchical agglomerative clustering methodology, the increase in the number of clusters improves the accuracy. The drawback of increasing the number of clusters is that the number of potential solutions subject to MC analysis also increases.

The adoption of a clustering algorithm with a non-crisp degree of membership, like the FCM algorithm, allows using the variable degree of membership as a measure of similarity between each potential solution to every cluster representative individual subject to MC analysis. The selection of the cluster representative individuals as the potential solutions with the best objective(s) value(s) per cluster assures that the most likely potential solutions to become part of the POF have their yield value accurately estimated. Then, by using a Euclidian-based degree of membership function it is possible to estimate the yield of the rest of potential solutions based on the distance to the cluster representative individuals, resulting in a different yield-estimated value to each potential solution.

The new fuzzy c-means based yield estimation (FUZYE) methodology clusters in the design parameter space all feasible potential solutions and select from each cluster a representative individual to perform the MC analysis. The representative individuals have their yield value accurately estimated, while for the rest of potential solutions of the population that were not subject to MC analysis have the yield estimated using the degree of membership and the cluster representative individuals yield value according to (4.41):

$$\widehat{Y}_{x_j} = \sum_{i=1}^{k} u_{ij} \cdot \widehat{Y}_{\text{RI}_i} \qquad (4.41)$$

where u_{ij} is the degree of membership of potential solution j to cluster i and $\widehat{Y}_{\text{RI}_i}$ is the estimated yield value of the cluster i representative individual.

The FUZYE methodology, Fig. 4.37, has four key steps: (1) identify the feasible individuals; (2) clustering of the feasible potential solutions; (3) selection of the representative individual from each cluster; and (4) assigning yield value to the remaining individuals in each cluster; as described in the following subsections.

4.3.6.1   First Step: Identify the Feasible Individuals

The FUZYE methodology first step, Fig. 4.38, implements an ISE process as described in Sect. 4.3.1. This step identifies the feasible potential solutions to avoid performing computational expensive MC simulations on solutions that do not comply with the circuit requirements under typical conditions. The ISE is achieved by dividing the evaluation process into two phases. The first phase evaluates all the individuals for typical conditions to obtain the circuit performance measures, and based on those values the solutions are classified as feasible or infeasible.

Feasible solutions adopt the new yield estimation technique, while for the infeasible solutions a negative yield value is assigned instead. The negative value is proportional to the amount of constraint violations by the solutions according to (4.32). For infeasible potential solutions, the evaluation process ends at this step.

4.3.6.2   Second Step: Clustering of the Feasible Potential Solutions

The FUZYE second step, Fig. 4.39, performs the clustering process of the feasible potential solution. The FUZYE methodology presented in this work adopts the FCM clustering algorithm as described in Sect. 4.2.3. The clustering algorithm implemented at this step follows the typical FCM clustering algorithm using as cluster representative individuals each cluster centroid, the selection of the best objective(s) value(s) individuals per cluster as cluster representatives is made in

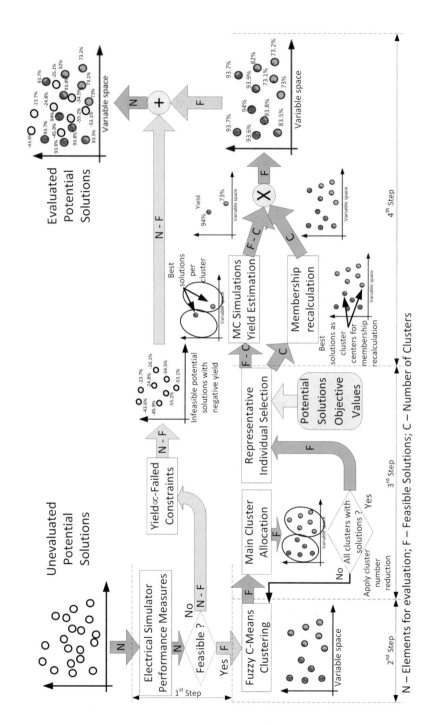

Fig. 4.37 FUZYE methodology flow

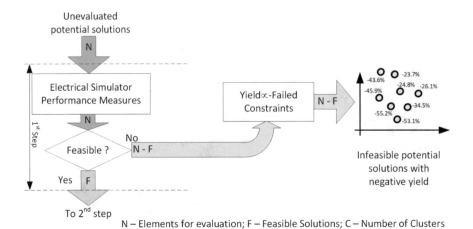

N – Elements for evaluation; F – Feasible Solutions; C – Number of Clusters

Fig. 4.38 FUZYE first step—infeasible solutions elimination

Fig. 4.39 FUZYE second
step—Clustering of feasible
potential solutions

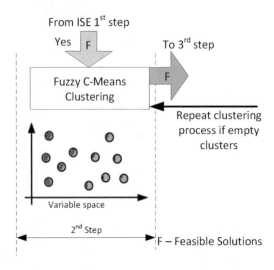

the next step of the FUZYE methodology. The implemented FCM algorithm defines
the degree of membership according to the squared Euclidian distance as in (4.12).

The graphic in Fig. 4.40 depicts the degree of membership function defined on the
real line for an example with three clusters where the cluster representative individuals are at positions $x = 1$, $x = 2$ and $x = 3$. The degree of membership graphic
shows that in case a potential solution is the cluster representative individual, the
function reaches its maximum value with respect to the cluster it represents and the
minimum for the other clusters, according to (4.13).

As was referred before, performing the FCM clustering algorithm requires the
definition of several parameters. Among the FCM algorithm parameters that require

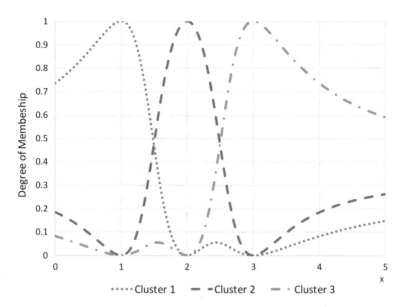

Fig. 4.40 Euclidian-based degree of membership function for 3 clusters example

setting is the number of clusters. One of the most widely approaches to select the number of clusters used in a clustering algorithm is the Elbow method. Based on the Elbow technique and the cost function (4.11), a method to automatically find the number of clusters at each iteration of the optimization algorithm was implemented as outlined in Algorithm 4.6. The implemented cluster number definition method performs the FCM algorithm using an increasing number of clusters and stops when the relative variation in the cost functions is below a small ε value.

Algorithm 4.6 Number of Clusters Definition

Input parameters:
    x – Feasible potential solution;

1	$k \leftarrow 1$, $N = \#Feasible\ potential\ solutions$
2	$J_m^{(1)} \leftarrow$ FCM$(\mathbf{x}, k)$ // FCM$(\cdot, \cdot)$fuzzy c-means according to Algorithm 4.3
3	While $(k < \sqrt{N})$
4	$\quad k \leftarrow k + 1$
5	$\quad J_m^{(k)} \leftarrow$ FCM$(\mathbf{x}, k)$
6	$\quad$ If $\left\| J_m^{(k)} - J_m^{(k-1)} \right\| / J_m^{(1)} < \varepsilon, 0 < \varepsilon < 1.$
	$\quad$ Then Exit while loop.

In the performed tests ε was set to 0.05. The total number of iterations was limited to $\sqrt{N}$, where N is the number of feasible potential solutions for clustering. This limit is used by most researchers to avoid exhaustive search [47]. Also, as the algorithm to define the number of clusters is based on the variation of cost function (4.11), which is monotonically decreasing for an increasing number of clusters, a punishing function or criterion must be imposed to avoid setting a number of clusters close to the total number of potential solutions [48]. The fact that the number of feasible potential solutions at each iteration of the optimization algorithm is limited by the population size, which in the tested circuit sizing and optimization tool adopted has typical values between 64 and 512, allows using this type of method to find the number of clusters with a time impact negligible on the overall optimization process. Like in the previous presented clustering methodologies and due to the possible large difference of magnitudes among dimensions in the design variable space, normalization per dimension is performed according to (4.36), which avoids that a particular dimension dominates the Euclidian distance measure and the clustering results.

Another important FCM parameter, as was discussed in Sect. 4.2.3, is the fuzziness parameter m. In order to assess the best value for this algorithm parameter several tests were performed using different test circuit sizing and optimization problems and different values of m. Since for values of $m \to 1$ the FCM clustering algorithm converges to the KMS results, only values with $m \geq 2$ were considered during the tests.

As referred, partitional clustering algorithms are very sensitive to the initial cluster centers selection. The choice of different random initial center data points can affect convergence speed and cluster identification. The center initialization method implemented in the presented FUZYE methodology is based on the k-means++ method as described at Sect. 4.2.4.

4.3.6.3 Third Step: Selection of the Cluster Representative Individual

After clustering all feasible potential solutions, the third step of the FUZYE methodology, Fig. 4.41, selects the cluster representative individual of each cluster, which is later subject to MC analysis. Based on the different studies and tests performed for the KMS representative element selection, the FUZYE methodology

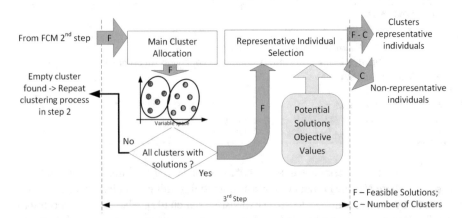

Fig. 4.41 FUZYE third step—Main cluster allocation and Cluster representative individual selection

adopts the technique presented in Sect. 4.3.2 for selecting the cluster representative individual. In this technique, the best individual from each cluster is identified and selected as the representative. The best individual per cluster is based on the objective (s) value(s) already calculated in the first evaluation phase at the first step of FUZYE. The cluster representative individual selection requires that every feasible potential solution be allocated to a cluster, and, since the FCM clustering algorithm does not have a crisp membership function like the KMS algorithm, a technique that allocates each potential solution to a specific cluster was developed. This technique examines the membership values for each potential solution and finds to which cluster the highest membership value occurs in order to allocate the potential solution to that cluster. The use of the best cluster membership value, for potential solutions allocation to clusters, may result in clusters that do not have any potential solution allocated, in which case the clustering process, implemented at step two of the FUZYE methodology, is repeated. For this repetition of the clustering process, a smaller number of clusters than in the previous clustering iteration is considered. The clustering process is repeated until all clusters has at least one potential solution allocated. The reduction in the number of clusters represents additional savings in expensive MC simulations, since fewer clusters mean fewer representative elements to simulate. The implemented cluster number reduction feature is later validated at Sect. 6.6.1.

4.3.6.4 Fourth Step: Assigning Yield Value to the Remaining Individuals in each Cluster

In the KMS methodology, the yield-estimated value for the remaining individuals in the cluster is identical to the value of the representative element. The non-crisp membership function expressed by the membership matrix, which measures the similitude between the representative elements to the rest of elements in the clusters, makes it possible to estimate the yield for the remaining individuals on the cluster in more detail than the KMS methodology. Matrix U with elements computed by (4.12), which is examined during the potential solutions cluster allocation process of the previous step, has the membership values based on the cluster centers calculated by the typical FCM algorithm, i.e., the cluster centroid. The selection of the best objective potential solutions as the representative elements, instead of the cluster centers, forces a membership adjustment considering the representative elements as the cluster centers (Fig. 4.42). The first step of this adjustment process is setting each cluster representative individual x_{RI_j} as a cluster center c_j (4.42),

$$c_i = x_{\mathrm{RI}_i}, \quad i = 1, \ldots, k \tag{4.42}$$

Once the new cluster centers are set, a single iteration updating the degree of membership of all clusters elements is performed. Since the representative individuals are subject to MC analysis, resulting in an accurate yield estimation value for

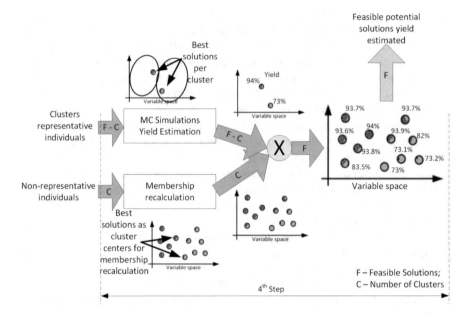

Fig. 4.42 FUZYE fourth step—MC analysis and yield estimation for non-simulated potential solutions

these particular potential solutions, the membership value for these individuals is set to 1 for the cluster that they represent and zero for the remaining clusters, according to:

$$
u_{ij} = \begin{cases} 1, & \text{if } x_j = c_i \\[2mm] \dfrac{1}{\sum_{p=1}^{k} \left(\dfrac{x_j - c_{i_2}^2}{x_j - c_{p_2}^2} \right)^{1/(m-1)}}, & \text{if } x_j \text{ is not a RI} \\[4mm] 0, & \text{otherwise} \end{cases} \tag{4.43}
$$

The degree of membership expressed in (4.43) results from combining (4.12) and (4.13). And, since the new cluster centers are the cluster representative individuals, the final degree of membership values reflects a similarity degree between each cluster elements to every MC-simulated cluster representative individual. Using the believe of similarity between non-simulated and simulated potential solutions, a yield-estimated value for every non-simulated potential solution is computed using (4.41). Since every potential solutions have its yield estimated based on the similarity to the simulated potential solutions, which according to the tests results improves the estimation accuracy when compared to the other presented cluster algorithms, the new yield estimation methodology based on FCM reduces the problem caused by projection of potential solutions yield into the yield line value of the cluster representative individual.

4.4 Conclusion

In this chapter, a new methodology for MC-based yield estimation using clustering algorithms to reduce the number of expensive MC simulations was presented. The reduction in the total number of MC simulations is crucial in order to include an accurate yield estimation technique, such as MC analysis yield estimation technique, in the loop of an evolutionary-based IC sizing and optimization tool.

The first clustering algorithm adopted in the new yield estimation methodology was the KMS algorithm. The original KMS algorithm suffered changes, namely in the selection of the cluster representative individual. The adopted cluster representative individual selection identifies at each cluster the solution with best objective (s) value(s), which assures that the most probable potential solutions to become part of the POF are going to be subject to MC simulations to accurately estimate its yield value. The KMS algorithm requires the previous definition for the number of clusters to perform the clustering task, to avoid specifying this algorithm parameter beforehand the HAC was tested. The HAC developed selects the number of clusters after performing the clustering task by using a method based on the data similarity, which resulted in a more accurate yield estimation than the KMS algorithm. The increase in yield estimation accuracy is achieved by performing a higher number of MC simulations. Despite the desirable increase in the yield estimation accuracy, the larger number of MC simulations is a major limitation when adopting the HAC algorithm in the optimization loop of an evolutionary-based IC sizing tool. To overcome the higher number of MC simulations by the HAC algorithm and increasing the yield estimation accuracy with respect to the KMS algorithm, the FCM algorithm was tested in the yield estimation methodology. Based on the FCM algorithm was developed the FUZYE methodology, which was able to achieve both goals, i.e., better yield estimation accuracy than KMS and fewer MC simulations than HAC algorithm. FUZYE adopts the degree of membership from the FCM algorithm to estimate the yield for the non-simulated potential solutions. The degree of membership is used as a measure of similitude between the simulated and non-simulated potential solutions at each generation of the optimization algorithm, allowing to estimate different yield values to each potential solution in the cluster, achieving more accurate yield estimation results than the KMS algorithm approach. The number of clusters setting technique and cluster number reduction feature implemented in the FUZYE methodology can reduce the total number of MC simulations when compared to the HAC algorithm, allowing the adoption of this new methodology in the analog IC sizing population-based optimization loop.

References

1. N. García-Pedrajas, J. Pérez-Rodríguez, Multi-selection of instances: a straightforward way to improve evolutionary instance selection. Appl. Soft Comput. **12**(11), 3590–3602 (2012)
2. C. Ding, X. He, K-means clustering via principal component analysis, in *Proc. 21st Int. Conf. Mach. Learn.*, Banff, Alberta, Canada, 2004

3. A. Fred, Similarity measures and clustering of string patterns, in *Pattern Recognition and String Matching*, (Springer, Boston, MA, 2003), pp. 155–193
4. S. Theodoridis, K. Koutroumbas, Chapter 11—Clustering: basic concepts, in *Pattern Recognition*, 4th edn., (Academic Press, Boston, MA, 2009), pp. 595–625
5. K.-L. Wu, J. Yu, M.-S. Yang, A novel fuzzy clustering algorithm based on a fuzzy scatter matrix with optimality tests. Pattern Recogn. Lett. **26**(3), 639–652 (2005)
6. T. Roughgarden, J.R. Wang, The complexity of the k-means method, in *24th Annual European Symposium on Algorithms (ESA 2016)*, Aarhus, Denmark, 2016
7. M. Masjed-Jamei, M.A. Jafari, H.M. Srivastava, Some applications of the stirling numbers of the first and second kind. J. Appl. Math. Comput. **47**(1), 153–174 (2015)
8. D. Steinley, K-means clustering: a half-century synthesis. Br. J. Math. Stat. Psychol. **59**(1), 1–34 (2006)
9. J. MacQueen, Some methods for classification and analysis of multivariate observations, in *Proc. 5th Berkeley Symp. Math. Stat. Probability*, 1967
10. S. Theodoridis, K. Koutroumbas, Chapter 5—Feature selection, in *Pattern Recognition*, 2nd edn., (Elsevier—Academic Press, San Diego, CA, 2003), pp. 163–205
11. C.D. Manning, P. Raghavan, H. Schütze, *Introduction to Information Retrieval* (Cambridge University Press, Cambridge, 2008)
12. L. Kaufman, P.J. Rousseeuw, *Finding Groups in Data: An Introduction to Cluster Analysis* (Wiley, New York, 1990)
13. R.T. Ng, J. Han, Efficient and effective clustering methods for spatial data mining, in *Proc. 20th Int. Conf. Very Large Data Bases (VLDB'94)*, Santiago de Chile, Chile, 1994
14. J.C. Bezdek, R. Ehrlich, W. Full, FCM: the fuzzy c-means clustering algorithm. Comput. Geosci. **10**(2), 191–203 (1984)
15. A. Stetco, X.-J. Zeng, J. Keane, Fuzzy C-means++. Expert Syst. Appl. **42**(21), 7541–7548 (2015)
16. C.H. Li, B.C. Kuo, C.T. Lin, LDA-based clustering algorithm and its application to an unsupervised feature extraction. IEEE Trans. Fuzzy Syst. **19**(1), 152–163 (2011)
17. M.-S. Yang, A survey of fuzzy clustering. Math. Comput. Model. **18**(11), 1–16 (1993)
18. L. Bai, J. Liang, C. Dang, F. Cao, A novel fuzzy clustering algorithm with between-cluster information for categorical data. Fuzzy Sets Syst. **215**, 55–73 (2013)
19. V. Schwämmle, O.N. Jensen, A simple and fast method to determine the parameters for fuzzy c–means cluster analysis. Bioinformatics **26**(22), 2841–2848 (2010)
20. V. Torra, On the selection of m for Fuzzy c-Means, in *2015 Conf. Int. Fuzzy Syst. Assoc. European Soc. Fuzzy Logic Technol. (IFSA-EUSFLAT-15)*, 2015
21. K.-L. Wu, Analysis of parameter selections for fuzzy c-means. Pattern Recogn. **45**(1), 407–415 (2012)
22. S. Ghosh, S.K. Dubey, Comparative analysis of K-means and fuzzy C-means algorithms. Int. J. Adv. Comput. Sci. Appl. **4**(4) (2013)
23. D.J. Ketchen, C.L. Shook, The application of cluster analysis in strategic management research: an analysis and critique. Strat. Manag. J. **17**, 441–458 (1996)
24. P.J. Rousseeuw, Silhouettes: a graphical aid to the interpretation and validation of cluster analysis. J. Comput. Appl. Math. **20**, 53–65 (1987)
25. D.T. Pham, S.S. Dimov, C.D. Nguyen, Selection of K in K-means clustering. Proc. Inst. Mech. Eng. C J. Mech. Eng. Sci. **219**(1), 103–119 (2005)
26. M. Halkidi, Y. Batistakis, M. Vazirgiannis, On clustering validation techniques. J. Intell. Inform. Syst. **17**(2), 107–145 (2001)
27. J.C. Bezdek, Numerical taxonomy with fuzzy sets. J. Math. Biol. **1**, 57–71 (1974)
28. J.C. Bezdek, Cluster validity with fuzzy sets. J. Cybernet. **3**, 58–74 (1974)
29. Y. Zhang, W. Wang, X. Zhang, Y. Li, A cluster validity index for fuzzy clustering. J. Inform. Sci. **178**(4), 1205–1218 (2008)
30. D. Campo, G. Stegmayer, D. Milone, A new index for clustering validation with overlapped clusters. Expert Syst. Appl. **64**, 549–556 (2016)

31. E. Lord, M. Willems, F.-J. Lapointe, V. Makarenkov, Using the stability of objects to determine the number of clusters in datasets. J. Inform. Sci. **393**, 29–46 (2017)
32. J. Wang, A linear assignment clustering algorithm based on the least similar cluster represen-tatives, in *Int. Conf. Syst. Man, Cybern.*, Orlando, FL, 1997
33. J. Fan, J. Wang, A two-phase fuzzy clustering algorithm based on neurodynamic optimization with its application for PolSAR image segmentation. IEEE Trans. Fuzzy. Syst. **26**(1), 72–83 (2016). https://doi.org/10.1109/TFUZZ.2016.2637373
34. K.L. Cheng, J. Fan, J. Wang, A two-pass clustering algorithm based on linear assignment initialization and k-means method, in *5th Int. Symp. Commun., Control Signal Process.*, Rome, 2012
35. D. Arthur, S. Vassilvitskii, k-means++: the advantages of careful seeding, in *Proc. 18th Annu. ACM-SIAM Symp. Discrete Algorithms (SODA'07)*, New Orleans, Louisiana, 2007
36. M.E. Celebi, H.A. Kingravi, P.A. Vela, A comparative study of efficient initialization methods for the k-means clustering algorithm. Expert Syst. Appl. **40**(1), 200–210 (2013)
37. A.K. Jain, M.N. Murty, P.J. Flynn, Data clustering: a review. ACM Comput. Surv. **31**(3), 264–323 (1999)
38. K. Abirami, P. Mayilvahanan, Performance analysis of K-means and bisecting K-means algorithms in weblog data. Int. J. Emerg. Technol. Eng. Res. **4**(8), 119–124 (2016)
39. R.R. Patil, A. Khan, Bisecting K-means for clustering web log data. Int. J. Comput. Appl. **116** (19), 36–41 (2015)
40. P. Cimiano, A. Hotho, S. Staab, Comparing conceptual, partitional and agglomerative cluster-ing for learning taxonomies from text, in *Proc. 16th European Conf. Artificial Intell.*, Amster-dam, 2004
41. L. Sousa, J. Gama, The application of hierarchical clustering algorithms for recognition using biometrics of the hand. Int. J. Adv. Eng. Res. Sci. **1**(7), 14–24 (2014)
42. F. Murtagh, P. Contreras, Algorithms for hierarchical clustering: an overview. WIREs Data Mining Knowl. Discov. **2**(1), 86–97 (2012)
43. G.W. Milligan, M.C. Cooper, An examination of procedures for determining the number of clusters in a data set. Psychometrika **50**(2), 159–179 (1985)
44. Y. Jung, H. Park, D.-Z. Du, B.L. Drake, A decision criterion for the optimal number of clusters in hierarchical clustering. J. Glob. Optim. **25**(1), 91–111 (2003)
45. R. Jenssen, D. Erdogmus, K.E. Hild, J.C. Principe, T. Eltoft, Information force clustering using directed trees, in *Energy Minimization Methods in Computer Vision and Pattern Recognition*, Lisbon, 2003.
46. G. Karypis, E.-H. Han, V. Kumar, Chameleon: hierarchical clustering using dynamic modeling. Computer **32**(8), 68–75 (1999)
47. M. Ren, P. Liu, Z. Wang, J. Yi, A self-adaptive fuzzy c-means algorithm for determining the optimal number of clusters. Comput. Intell. Neurosci. **2016**, 12 (2016)
48. X.L. Xie, G. Beni, A validity measure for fuzzy clustering. IEEE Trans. Pattern Anal. Mach. Intell. **13**(8), 841–847 (1991)

Chapter 5
AIDA-C Variation-Aware Circuit Synthesis Tool

5.1 AIDA-C Analog IC Design Flow

AIDA-C [1] analog IC sizing tool is part of the AIDA framework which also includes AIDA-L [2, 3], an analog IC layout automation design tool. AIDA-C adopts the HAD methodology at circuit level as described in Sect. 2.2. AIDA-C sizing tool was developed in Java™ 1.6 with a fully graphical user interface, Fig. 5.1. Since all circuit sizing problems presented in this book were solved using Mentor Graphics' ELDO™ [4] electrical simulator, the implementation details discussed in the next sections refer to this electrical simulator.

AIDA-C block diagram is shown in Fig. 5.2. Two main blocks stand out from AIDA-C, the Setup and Monitoring and the Multi-objective Optimizer. The Setup and Monitoring block modules, with a user-friendly graphical user interface (GUI), help designers defining the circuit to be optimized as an optimization problem and allow designers keep up with developments of the optimization process. The Multi-objective Optimizer block module implements a multi-objective multi-constraint simulation-based optimization kernel that takes the circuit sizing and optimization problem defined at the Setup module and applies swarm-base and/or population-based optimization techniques to solve the circuit sizing problem.

5.1.1 Setup and Monitoring Block Modules

AIDA-C Setup module requires a circuit representation as input. The circuit representation is supplied by IC designers in the form of a parameterized circuit netlist file. The netlist file can be produced by hand but is usually obtained by exporting the circuit from a schematic's editor, such as Virtuoso Schematic Editor from Cadence®. An example of a circuit schematic and circuit representation netlist is shown in Fig. 5.3.

© Springer Nature Switzerland AG 2020
A. M. L. Canelas et al., *Yield-Aware Analog IC Design and Optimization in Nanometer-scale Technologies*, https://doi.org/10.1007/978-3-030-41536-5_5

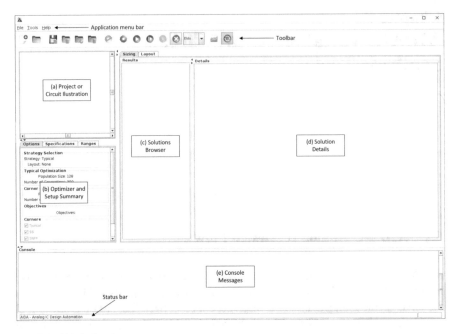

Fig. 5.1 AIDA-C main screen

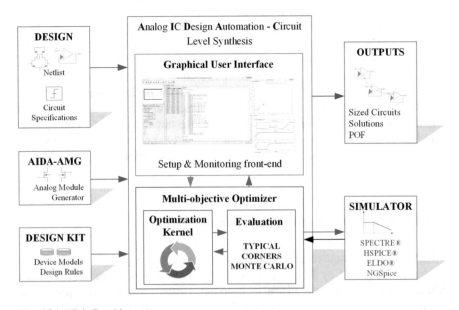

Fig. 5.2 AIDA-C architecture

a

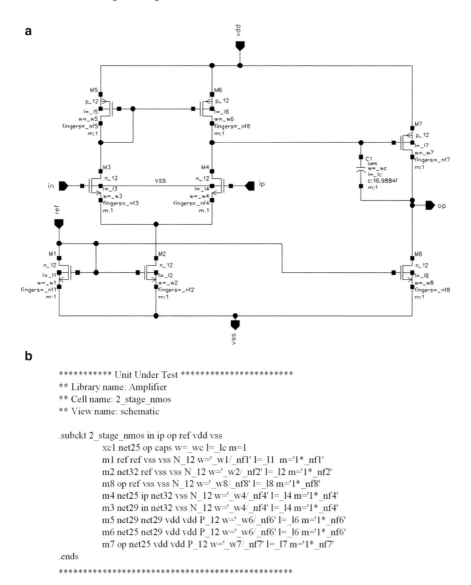

b

```
*********** Unit Under Test **********************
** Library name: Amplifier
** Cell name: 2_stage_nmos
** View name: schematic

.subckt 2_stage_nmos in ip op ref vdd vss
        xc1 net25 op caps w=_wc l=_lc m=1
        m1 ref ref vss vss N_12 w='_w1/_nf1' l=_l1  m='1*_nf1'
        m2 net32 ref vss vss N_12 w='_w2/_nf2' l=_l2 m='1*_nf2'
        m8 op ref vss vss N_12 w='_w8/_nf8' l=_l8 m='1*_nf8'
        m4 net25 ip net32 vss N_12 w='_w4/_nf4' l=_l4 m='1*_nf4'
        m3 net29 in net32 vss N_12 w='_w4/_nf4' l=_l4 m='1*_nf4'
        m5 net29 net29 vdd vdd P_12 w='_w6/_nf6' l=_l6 m='1*_nf6'
        m6 net25 net29 vdd vdd P_12 w='_w6/_nf6' l=_l6 m='1*_nf6'
        m7 op net25 vdd vdd P_12 w='_w7/_nf7' l=_l7 m='1*_nf7'
.ends

**************************************************
```

Fig. 5.3 (**a**) Example circuit schematic and respective (**b**) parameterized netlist

The netlist circuit representation section detailed at Fig. 5.3b corresponds to the circuit whose schematics is presented at Fig. 5.3a. Notice that instead of setting a numerical value to every device parameters, i.e., *w*, *l*, and *m*, to most of them is assigned a variable, which is used later during the optimization processes as a design variable.

The circuit netlist must include more information than just the circuit representation. Because AIDA-C optimization module invokes the electrical simulator using

```
*********** Test-bench *************************
** Library name: Amplifier
** Cell name: 2_stage_nmos_OLtb
** View name: schematic
xi0 in_n in_p output net17 net021 0 2_stage_nmos
vinn in_n 0 DC vcm
vinp in_p 0 DC vcm AC 1
vdd net021 0 DC vdd
rload output 0 resload
cload output 0 capload
ibias net021 net17 DC ibias
************************************************


*********** Analysis *************************
.TEMP 25.0
.AC DEC 20 1 10G SWEEP DATA = PIPEdata
.OPTION BRIEF=0
************************************************

*********** Performance Measures **************
.EXTRACT AC  label=GDC  YVAL(VDB(OUTPUT),1)
.EXTRACT AC  label=GBW  XDOWN(VDB(OUTPUT),0,start,end)
.EXTRACT AC  label=PM   180+MIN(VP(OUTPUT),start,Extract(GBW))
************************************************
```

Fig. 5.4 Netlist test-bench, circuit analysis, and performance measures sections example

the netlist file, in order to perform the required circuit simulations, it is necessary to include additional sections, such as a test-bench circuit section to extract the different circuit performance measures and a circuit analysis section and circuit performance measures required to compute the optimization problem constraints and objectives. Since more than one test-bench circuit may be required to obtain the different performance measures, several netlist files must be generated. In Fig. 5.4, a test-bench circuit, an AC analysis, and performance measures sections are shown; the presented netlist example is intended to be simulated by Mentor Graphics' ELDO™ electrical simulator.

Once the netlist file or files are all generated, the setup process in AIDA-C may start. The setup process starts by creating a new optimization project in AIDA-C main screen. This simple setup process opens an intuitive form, Fig. 5.5, which allows designers specifying the technology node adopted for the circuit to select the electrical simulator among the different simulators supported and set the netlist files with the different test-bench circuits to obtain the necessary circuit performance measures.

Before the new AIDA-C Variation-Aware version was developed, the only technique available in AIDA-C to improve circuit solution robustness was performing optimization considering worst-case process, voltage and temperature (PVT) corner analysis, which in today's technology nodes is considered insufficient [5]. Despite

Fig. 5.5 AIDA-C new project form

some of the corner analysis limitations, the fact is that it is a well-accepted technique by IC designers. Configuring corner analysis is also possible in the Setup form by defining the corners test-bench(es) to include in the sizing and optimization processes.

The "Auto-Setup" tab form in Fig. 5.5 allows designers to automatically add several predefined circuit functional constraints and the respective measures in the netlist/test-bench file to obtain their value, such as the overdrive voltage and minimal saturation margin, ensuring that CMOS transistors are in the saturation region. Whereas the "Variables" tab is where IC designers must specify the circuit sizing design variables that define the optimization problem search space. The variables are defined as range variables, where a minimum and maximum value must be set, and a grid step for each variable must also be set. The Setup process creates an extensible markup language (XML) file where all setup information is saved. The setup file, named *design.xml*, is accessible for later edition by designers.

Finished this first step of the setup process, by filling Fig. 5.5 form, only the problem optimization variables and how to compute their values is setup. In order to conclude the optimization problem setup process, the problem constrains and optimization objectives must be defined. This last task of the setup process is achieved at the AIDA-C setting screen, where it is possible to select the objectives among the several defined circuit performance measures and, also, define the goal for each objective, i.e., maximization or minimization. At the same screen, the different problem constraints are defined by selecting the measure, setting the constraint boundary value and selecting the constraint condition sign, i.e., *greater than or equal to* ($\geq$) or *less than or equal to* ($\leq$). Since only two constraints conditions are available ($\geq$ and $\leq$), designers must add two constraints as defined in (2.6) to set up an equality constraint condition. After performing these two simple setup steps, the IC sizing problem is now converted to an optimization problem that AIDA-C Multi-Objective Optimizer block can solve.

The optimization Monitoring module supplies information about the evolution of the optimization processes. Since IC sizing and optimization is typically a multi-objective problem, the Monitoring module updates at each iteration or generation of the optimization algorithm a graphical representation of the current Pareto front. In Fig. 5.6a the solution space considering the problem objectives is represented, the nondominated solutions are presented as dark green circles (dark gray in monochrome print), while the best non-feasible solutions in early stages of the optimization process are plotted as different symbols in the top section of the graphic. The Monitoring module can represent up to three-dimensional Pareto solutions in a single graphical window. For more than three objectives, multiple graphical windows are shown, each representing different projections of the Pareto solutions by combining pairs of objectives. Because the Multi-Objective Optimizer module only solves minimization problems, all axes in Fig. 5.6a present negative values, as maximization problems are converted into minimization optimization problems by multiplying the objective measure by -1.

In addition to presenting the evolution of Pareto solutions, the Monitoring module also presents four valuable indicator graphics that supply information about the convergence of the current optimization processes, Fig. 5.6b. The four indicators are from top to bottom: a graphic detailing the number of optimal nondominated solution vs. the generation number of the algorithm; next, the *gsum* indicator is shown, which reflects by how far the best infeasible solutions is from feasibility and is computed as the sum of the failed constraints normalized distance to feasibility (5.1).

$$gsum = \sum_{j=1}^{\#\text{Constraints}} \begin{cases} \dfrac{G_j - g_j(\mathbf{x_d})}{G_j + \varepsilon} & \text{if } g_j(\mathbf{x_d}) > G_j \\ 0 & \text{otherwise} \end{cases} \tag{5.1}$$

where G_j is the boundary value of constraint j, $(g_j(\mathbf{x_d}) \leq G_j)$.

According to (5.1), after reaching the feasibility this indicator becomes zero; the third graphic presents the dominated area, which is generally the relative hyper-volume of the region dominated by the current Pareto front, and gives an important information about the evolution of the dominated area, which allows designers to stop the optimization process when no improvements are related for several generations; finally, the number of feasible offspring's at each generation is detailed.

a

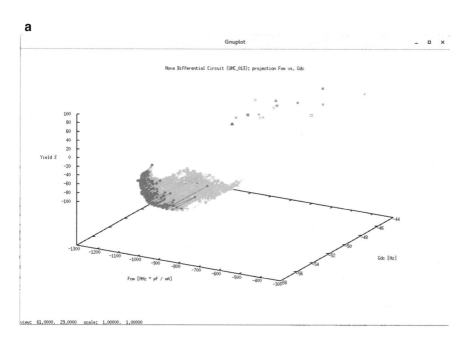

b

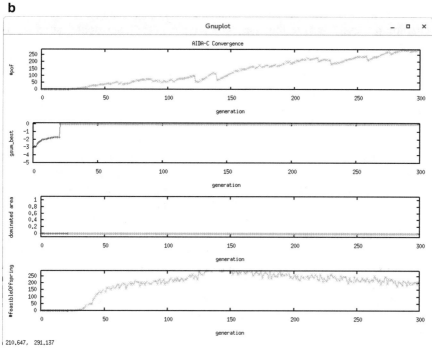

Fig. 5.6 AIDA-C Monitoring module information. (**a**) Pareto solutions evolution. (**b**) Optimization convergence

5.1.2   Multi-Objective Optimizer Block

AIDA-C Multi-Objective Optimizer block is built around two main modules: the optimization kernel and the evaluator module. The optimization kernel implements optimization techniques such as NSGA-II, multi-objective simulated annealing (MOSA) [6], and MOPSO [7]. The adopted multi-objective optimization techniques are based on a large number of potential solutions that explores the variable space of the optimization problem to guide the search for the optimal solutions. Although several metaheuristic optimization techniques are available at AIDA-C optimization kernel, tests performed using circuit sizing and optimization problems revealed that the NSGA-II-based optimization kernel was able to attain the best results in terms of solutions reached [8]. Accordingly, to the good results obtained by the NSGA-II algorithm, all production software releases of AIDA-C were developed focusing the NSGA-II optimization algorithm. Although AIDA-C NSGA-II default parameter values are set to achieve good optimization results at most IC sizing and optimization problems, it is possible setting other values if the sizing problem so dictates. In the typical optimization parameters setting window can be set the following optimization parameters: population size; maximum number of generations, which is adopted as the stopping criteria for the optimization processes; mutation rate; and finally the crossover rate, Fig. 5.7.

Since AIDA-C offers the feature of automatically performing a two-phase optimization process, where in the first phase a typical conditions optimization run is performed followed by a second phase PVT corners optimization process, a similar optimization parameters setting screen is available for the corners optimization phase, allowing different settings at each optimization phase, Fig. 5.8.

The evaluator module invokes the selected electrical simulator to compute the optimization problem objectives and constraints for each population individual. According to the type of optimization evaluation, i.e., typical or corners, Fig. 5.7, the evaluator defines the test-bench file to perform the circuit simulation. Since several problem constraints and/or objectives do not depend on electrical circuit measures, like the circuit area, or require additional computation based on the measures returned by the electrical simulator, the evaluator module must perform these additional computations. The estimated circuit area is accurately computed by using the analog module generator [9] (AMG). AIDA-AMG can instantiate real circuit modules, depending on the selected technology, going from simple capacitors or transistor structures to more complex structures such as folded transistors, merged transistors, interdigitized structures, and common-centroid pattern transistor structures, Fig. 5.9. The modules are laid on the circuit floorplan, based on a predefined layout template, which gives an accurate value for the total circuit area, without performing a complete circuit layout, but where all technology design rules are respected.

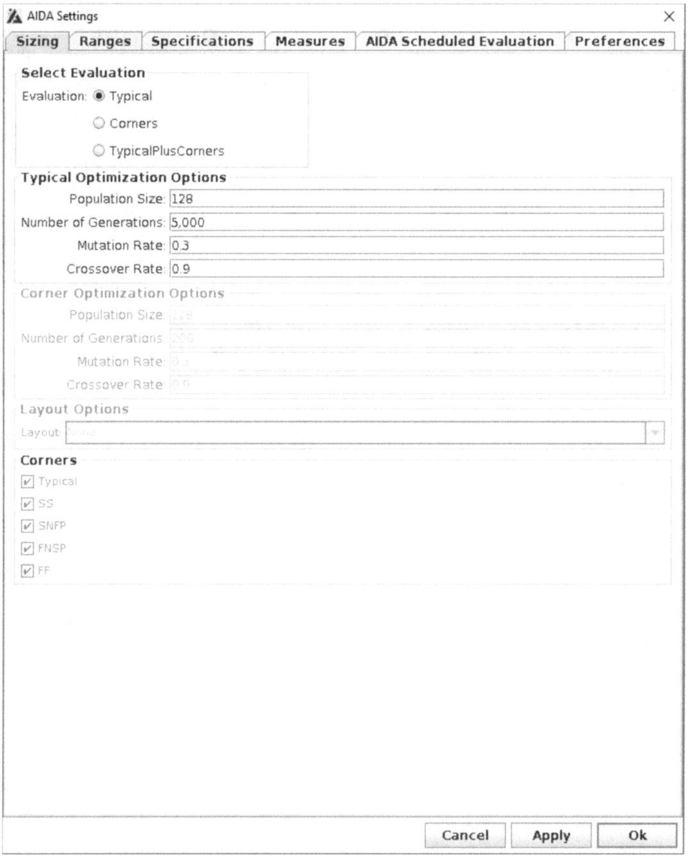

Fig. 5.7 AIDA-C optimization setting parameters

Fig. 5.8 AIDA-C evolutionary parameters setting form

a   **b**   **c**   **d**

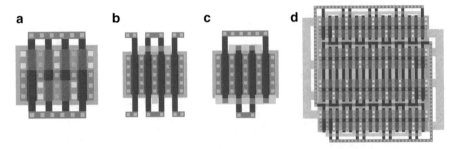

Fig. 5.9 AIDA-AMG transistor structures examples. (**a**) Folded structure for a 4-fingers transistor. (**b**) Merged structure for 3 transistors with unconnected gates. (**c**) Interdigitized structure for 2 transistors with 2 fingers. (**d**) Common-centroid structure for 2 transistors with 32 fingers

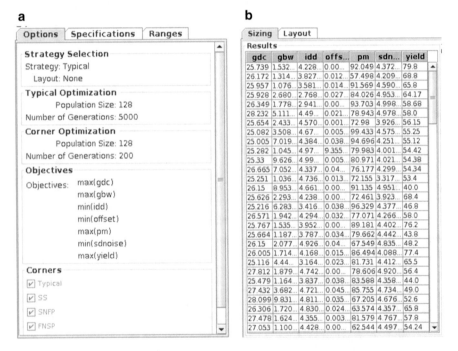

Fig. 5.10 AIDA-C main panels. (**a**) Optimizer and setup summary panel. (**b**) Solutions browser panel

5.2   AIDA-C Main Graphical User Interface

AIDA-C main GUI screen, Fig. 5.1, is the program front-end where all AIDA-C modules interact with users/designers and is divided into five main panels. In the project or circuit illustration panel an image, which the designer can specify in the design.xml setup file, is shown. The image allows to easily identify the circuit under

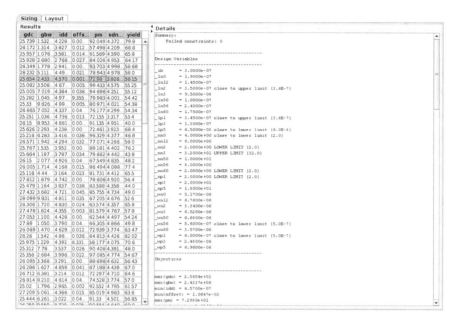

Fig. 5.11 AIDA-C solution detail panel

optimization when several AIDA-C instances, with different circuits, are running at the same time. The Optimizer and Setup Summary panel, Fig. 5.10a, presents information related to the optimization algorithm settings, circuit specifications, and optimization variable ranges. The information is divided among three tabs identified as "Options," "Specifications," and "Ranges", respectively. The Solutions Browser panel, Fig. 5.10b, presents nondominated solutions achieved so far by the current optimization processes, detailing the objective(s) value(s) achieved by each solution.

Since at the beginning of the optimization processes, typically, all population individuals are infeasible, the solutions browser only presents the individual close to the feasibility region based on the *gsum* parameter (5.1). After selecting a particular solution from the solution browser, the Solution Detail panel, Fig. 5.11, presents the design variables values, objectives values, performance constraints values, and the electrical measures return by the electrical simulator for the selected solution. The Console Messages panel presents different status messages about the performed AIDA-C tasks.

At the solution browser panel, an additional context-menu is opened by a mouse right-button click on a particular solution, Fig. 5.12. The context-menu allows designers to include new solutions or edit existing ones, deleting solutions is also possible. Using this menu makes possible to designers simulate the selected solution and export the solution into AIDA-L, to obtain an optimized layout based on the selected sized solution.

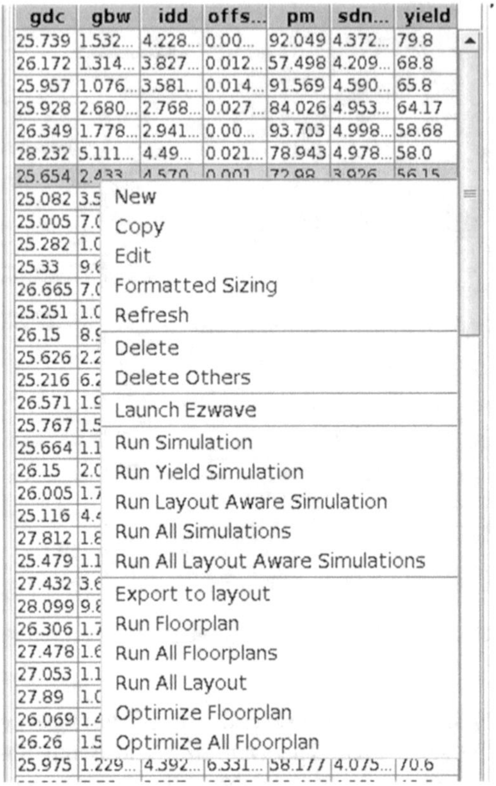

Fig. 5.12 Solutions
browser context-menu

gdc	gbw	idd	offs...	pm	sdn...	yield
25.739	1.532...	4.228...	0.00...	92.049	4.372...	79.8
26.172	1.314...	3.827...	0.012...	57.498	4.209...	68.8
25.957	1.076...	3.581...	0.014...	91.569	4.590...	65.8
25.928	2.680...	2.768...	0.027...	84.026	4.953...	64.17
26.349	1.778...	2.941...	0.00...	93.703	4.998...	58.68
28.232	5.111...	4.49...	0.021...	78.943	4.978...	58.0
25.654	2.433...	4.570	0.001	72.98	3.926	56.15
25.082	3.5	New				
25.005	7.0	Copy				
25.282	1.0	Edit				
25.33	9.6					
26.665	7.0	Formatted Sizing				
25.251	1.0	Refresh				
26.15	8.9	Delete				
25.626	2.2					
25.216	6.2	Delete Others				
26.571	1.9	Launch Ezwave				
25.767	1.5					
25.664	1.1	Run Simulation				
26.15	2.0	Run Yield Simulation				
26.005	1.7	Run Layout Aware Simulation				
25.116	4.4					
27.812	1.8	Run All Simulations				
25.479	1.1	Run All Layout Aware Simulations				
27.432	3.6	Export to layout				
28.099	9.8					
26.306	1.7	Run Floorplan				
27.478	1.6	Run All Floorplans				
27.053	1.1	Run All Layout				
27.89	1.0					
26.069	1.4	Optimize Floorplan				
26.26	1.5	Optimize All Floorplan				
25.975	1.229...	4.392...	6.331...	58.177	4.075...	70.6

AIDA-C toolbar, Fig. 5.1, presents a set of buttons that work as shortcuts to the most common tasks. From left to right, the first six buttons are related to file operations that allow designers to create a new project setup file, open an existing project, save the project settings and current solutions, load settings and solutions from a previous optimization run, load the project settings, and, finally, load solutions from a previous run. The next two buttons are related to project setting tasks. The first button, in this setting group buttons, performs the validation of the current project settings, which includes validating the netlist/test-bench(es) file(s) by the electrical simulator. The second button opens the AIDA-C settings window, Fig. 5.13, whose first "Sizing" tab was already presented in Fig. 5.7.

The "Ranges" tab, at AIDA-C settings window, presents the optimization problem variables, previously set at the setup process or in the *design.xml* setup file, and respective ranges. Notice that this window does not allow adding new variables; it is only possible to edit the ranges or enable/disable a variable. In order to add new variables, designers must edit the *design.xml* setup file and manually add the desired variables.

AIDA-C settings window Specifications tab presents the problem objectives and constraints in a tree view graphical control, Fig. 5.14, which facilitates project maintenance. The problem constraints appear under the *Specifications* tree view

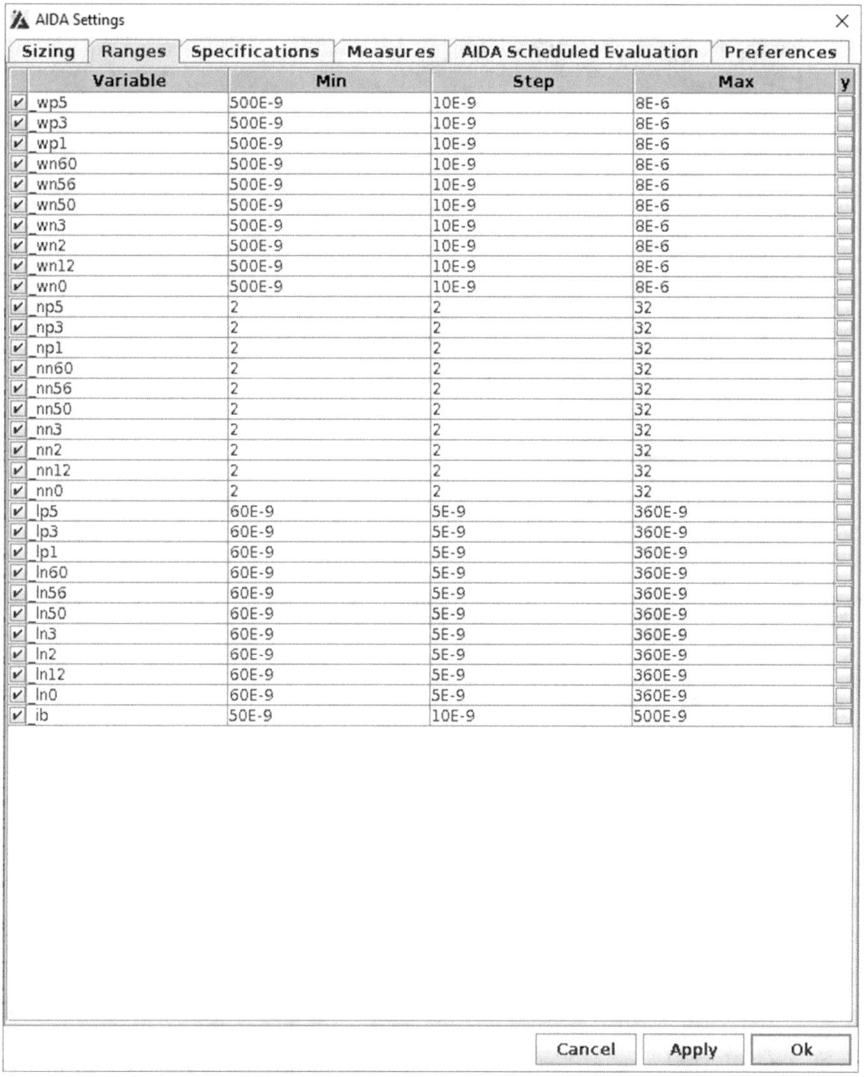

Variable	Min	Step	Max	y
☑ wp5	500E-9	10E-9	8E-6	☐
☑ wp3	500E-9	10E-9	8E-6	☐
☑ wp1	500E-9	10E-9	8E-6	☐
☑ wn60	500E-9	10E-9	8E-6	☐
☑ wn56	500E-9	10E-9	8E-6	☐
☑ wn50	500E-9	10E-9	8E-6	☐
☑ wn3	500E-9	10E-9	8E-6	☐
☑ wn2	500E-9	10E-9	8E-6	☐
☑ wn12	500E-9	10E-9	8E-6	☐
☑ wn0	500E-9	10E-9	8E-6	☐
☑ np5	2	2	32	☐
☑ np3	2	2	32	☐
☑ np1	2	2	32	☐
☑ nn60	2	2	32	☐
☑ nn56	2	2	32	☐
☑ nn50	2	2	32	☐
☑ nn3	2	2	32	☐
☑ nn2	2	2	32	☐
☑ nn12	2	2	32	☐
☑ nn0	2	2	32	☐
☑ lp5	60E-9	5E-9	360E-9	☐
☑ lp3	60E-9	5E-9	360E-9	☐
☑ lp1	60E-9	5E-9	360E-9	☐
☑ ln60	60E-9	5E-9	360E-9	☐
☑ ln56	60E-9	5E-9	360E-9	☐
☑ ln50	60E-9	5E-9	360E-9	☐
☑ ln3	60E-9	5E-9	360E-9	☐
☑ ln2	60E-9	5E-9	360E-9	☐
☑ ln12	60E-9	5E-9	360E-9	☐
☑ ln0	60E-9	5E-9	360E-9	☐
☑ ib	50E-9	10E-9	500E-9	☐

Fig. 5.13 AIDA-C settings window

top element in a hierarchical structure where it is possible to define if a constraint is applied only to typical and/or corners conditions optimization. Unlike the Ranges tab where it is not possible to add new variables, at the "Specifications" tab designers can delete, edit, or add objectives or constraints by using the context-menu, available by a mouse right-click. This context-menu offers the ability of enabling or disabling objectives or constraints and check their properties, Fig. 5.15.

The objective properties window, Fig. 5.15a, allows editing an existing objective. At this window designers can change the measure and define the objective

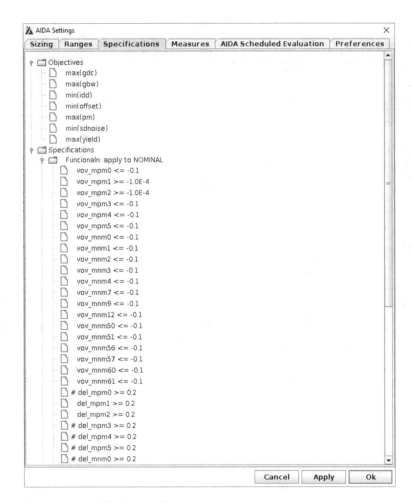

Fig. 5.14 AIDA-C specifications settings screen

goal, i.e., *Maximize* or *Minimize*. The constraint properties window, Fig. 5.15b, allows selecting the circuit measure, defining the type of constraint, and setting the boundary value for the constraint.

The next group of buttons in the toolbar, Fig. 5.16, using typical images from a music player device or application, are related to the optimization processes.

The play and step-forward buttons start the optimization processes, whereas the stop button interrupts the current optimization processes. The delete button, with a big white cross, deletes all solutions/individuals from the population, in order to start the new optimization process with a new random population. The combo box in the toolbar allows selecting the electrical simulator, Fig. 5.17. Finally, the last two buttons in the toolbar, Fig. 5.17, are used to enable or disable the monitoring graphic

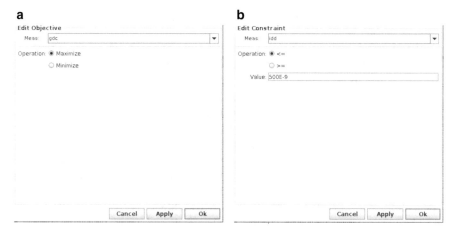

Fig. 5.15   AIDA-C (**a**) objectives and (**b**) constraints properties windows

Fig. 5.16   AIDA-C
optimization processes
toolbar

Fig. 5.17   AIDA-C
electrical simulator selector
and monitoring windows
control

windows, Fig. 5.6, showing the POF and the optimization algorithm convergence information.

AIDA-C offers a user-friendly project setup process and GUI that provides real online information about the evolution of the optimization process. It allows designers to actuate on the optimization process, either by changing some optimization parameters, like the size of the population, or by redefining some active constraints in order to faster obtain solutions.

5.3   AIDA-C Variation-Aware Implementation

This section describes the implementation of the new yield-aware feature of AIDA-C. Incorporating in AIDA-C the new yield estimation methodologies, described at Chap. 4, followed the approach of disturbing as minimum as possible the already stable modules that constitute AIDA-C. The new feature required changes, mainly, at the evaluation module and GUI, and, of course, some minor changes at the project setup procedure.

The electrical simulator requires a netlist and test-bench file where instead of typical or corners device models the variability foundry device models are used, and the commands to perform MC simulations are specified. So, during the AIDA-C setup process, designers must create an additional netlist and test-bench file for MC purposes. The step of creating this particular MC netlist file is quite easy, if the typical netlist file was already created, since the new file is almost identical to the typical netlist file. In fact, a copy of the typical/corners netlist file using the same name with the prefix "mc_" is required. Once the copy is made, the designer must edit the netlist file to include the variability device models and introduce the commands to perform the MC simulations. Because of the adopted file name convention, the specification of the MC netlist file is not required during the AIDA-C setup process, since AIDA-C will search and adopt as MC netlist file (s) the netlist file(s) with prefix "mc_" and with the rest of the name identical to the typical or corners netlist file(s), according to the optimization conditions selected.

The previous evaluator module of AIDA-C already invokes the electrical simulator for electrical performance measures estimation to evaluate the algorithm population and return the results to the optimization kernel. So, the new second evaluation phase, required by the new yield estimation methodology, was inserted in the previous evaluation flow after results are returned by the electrical simulator. In Fig. 5.18, the major additions to the evaluator module are identified by the shaded area. The added functionalities to the evaluation module start by reading the simulation results and computing the *gsum* parameter (5.1). Based on this parameter, solutions are classified as feasible (*gsum* = 0) or infeasible (*gsum* < 0), step (1) of Fig. 5.18. Next, a new procedure in the evaluation module receives a list of infeasible solutions and a negative yield value is computed based on (4.32), step (3). At step (2), feasible solutions are clustered by the selected clustering algorithm and cluster representative individuals are selected. Next, the selected electrical simulator is invoked using as parameters a file with the cluster representative individuals and the file with the MC simulations test-bench(es). The selected electrical simulator returns the results of the MC simulations in a single file where for each cluster representative individual the different MC iterations performed by the electrical simulator are detailed. The use of the new yield estimation methodology with PVT corners is also possible, when using this optimization mode, the results file returns the different corners simulation and for each corner and circuit solution the results of the performed MC iterations by the electrical simulator. The MC terminology adopted by this document is based on the terms used by the selected electrical simulator, i.e., Mentor Graphics' ELDO™. This simulator refers to MC simulation as the set of simulations performed on each circuit solution considering variability PDK device models and a MC iteration is one of the simulations performed during a MC simulation request. So, for each MC simulation it is possible to perform a desired number of iterations to attain the required results yield estimation accuracy.

Since, depending on the optimization conditions considered (*Typical* + *MC* or *Corners* + *MC*), the simulations result files formats are different, two new parsers

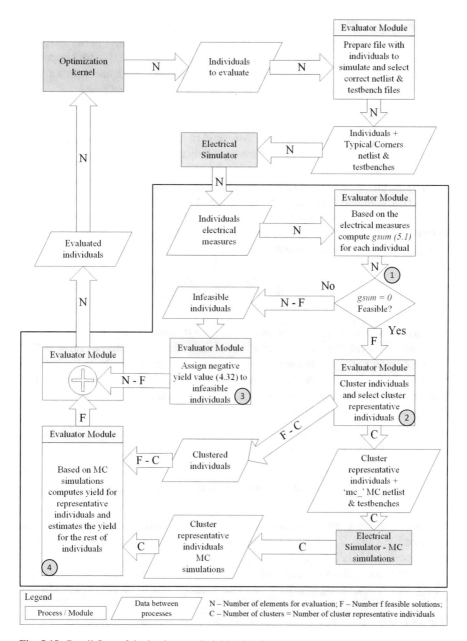

Fig. 5.18 Detail flow of the implemented yield estimation methodology

were developed. The *Typical + MC* new parser reads for each simulated cluster representative every iterations result, whereas the *Corners + MC* parser reads for each simulated cluster representative and each corner the respective MC iterations. As AIDA-C considers the yield as a performance measure, each corner has a yield value, and for optimization purposes the worst-case value is considered.

Both parsers invoke a module to compute the yield of each simulated individual and determine their respective process capability index or C_{pk} values for each performance measure constraint. The final yield measure for a cluster representative individual is computed using (3.5), where the indicator function returns the number of MC iterations where all optimization problem constraints are fulfilled, and N refers to the number of MC iterations performed. The process capability index (5.2) [10], computed for each cluster representative and performance measure constraint, provides information about each performance measure constraint by measuring how far is the average of the performance constraint samples from the boundary limit constraint value in terms of 3σ.

$$C_{pk}^{(j)} = \min\left(\frac{\mu - G_L^{(j)}}{3\sigma}, \frac{G_U^{(j)} - \mu}{3\sigma}\right) \qquad (5.2)$$

where μ is the average performance value estimated using the different MC iterations; also, using the MC iterations is computed the standard deviation σ. The G's terms refer to the j constraint boundary condition values, which are the imposed target performance measures, specified as problem constraints $G_L \leq g_j(\mathbf{x}_d) \leq G_U$. Since most problem constraints are only one-sided constraints [11], on those cases the C_{pk} is computed by (5.3):

$$C_{pk}^{(j)} = \begin{cases} \dfrac{\mu - G_L^{(j)}}{3\sigma}, & \text{if constraint } j \text{ is of the form}: \ G_L \leq g_j(\mathbf{x}_d) \\[2ex] \dfrac{G_U^{(j)} - \mu}{3\sigma}, & \text{if constraint } j \text{ is of the form}: \ g_j(\mathbf{x}_d) \leq G_U \end{cases} \qquad (5.3)$$

Since a relation exists between the standard deviation σ and the yield, as discussed in Chap. 2, it is also possible to relate the process capability index with the process yield [12], Table 5.1.

The process capability index supplies a very important information to designers. Using this index, designers have a much clear information about performance

Table 5.1 Relation between process capability index and yield for two-sided constraints	C_{pk}	Yield (%)
	1.0	99.7300204
	1.2	99.9681783
	1.4	99.9973309
	1.6	99.9998413
	1.8	99.9999933
	2.0	99.9999998

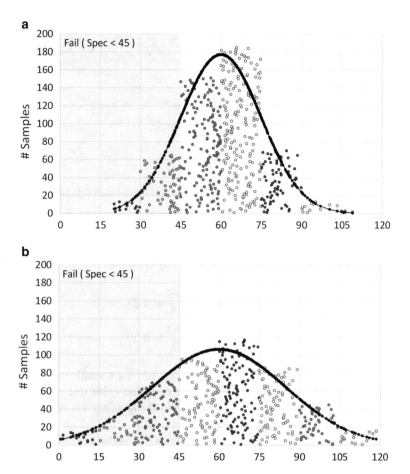

Fig. 5.19 Two MC iteration data sets histogram of 500 unitless samples with identical mean ($\mu = 60$) and different standard deviations (**a**) $\sigma = 15$, (**b**) $\sigma = 25$

measures constraints that have a major impact on the final yield value, since lower C_{pk} values indicate that the respective measure constraint is more difficult to achieve. A lower C_{pk} value indicates constraints with smaller safety margins to the specification limits, because its sample mean value is close to the specification constraint boundary value considering the natural variability of the specification measure.

Another reason for lower C_{pk} values is related to the standard deviation of the MC specification samples used to compute process capability index. In Fig. 5.19, an example of two data sets of 500 MC iteration samples for the same unitless circuit specification constraint measure is depicted.

The example in Fig. 5.19 presents the same distance between the specification constraint boundary ($G_L = 45$) and the mean value of the samples ($\mu = 60$). The difference between data sets are the standard deviation values, $\sigma = 15$ and $\sigma = 25$ for

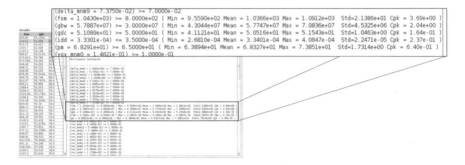

Fig. 5.20 New solution detail panel, including the performance constraints C_{pk} index and statistical information

Fig. 5.19a and Fig. 5.19b, respectively. Using the referred values, the process capability indexes for each data samples are easily computed:

$$C_{pk}^{\text{Fig.5.19a}} = \frac{60 - 45}{3 \times 15} = 0.33 \tag{5.4}$$

$$C_{pk}^{\text{Fig.5.19b}} = \frac{60 - 45}{3 \times 25} = 0.20 \tag{5.5}$$

As expected, the C_{pk} value from Fig. 5.19b in (5.5) has a smaller value than C_{pk} from Fig. 5.19a because of the higher standard deviation value that causes a larger data spread, which leads to more samples failing the specification constraint, thus lowering the yield of that particular specification.

Reporting the yield and process capability index to designers required changes at AIDA-C solutions detail panel. Along with the C_{pk} index was also added information about the performance measure minimum, mean, maximum, and standard deviation values of the MC iterations, Fig. 5.20. The different reported values are computed as the parser reads the results avoiding storing the complete MC iterations results in memory.

Several other changes were also implemented on different areas of the GUI. The new yield optimization technique, where two evaluation processes exists and the problem optimization constraints' condition for typical and yield may differ, led to some changes in the GUI of AIDA-C circuit sizing tool. As depicted in Fig. 5.18, the first evaluation process assesses the typical/corners performance measures based on the circuit specifications conditions detail in the AIDA-C settings "Specification" tab, Fig. 5.14. Since at the beginning of the optimization process IC designers may define constraints with a very tight criterion, which helps getting the GA algorithm to reach robust feasible solutions, during the yield evaluation and estimation process it is acceptable to relax or even disable some of these constraints. The effect of this relaxation is to focus the yield optimization, i.e., the solutions that comply the problem specifications, on the real important measures that must hold under variability conditions. An example of this procedure is to relax or completely disable the

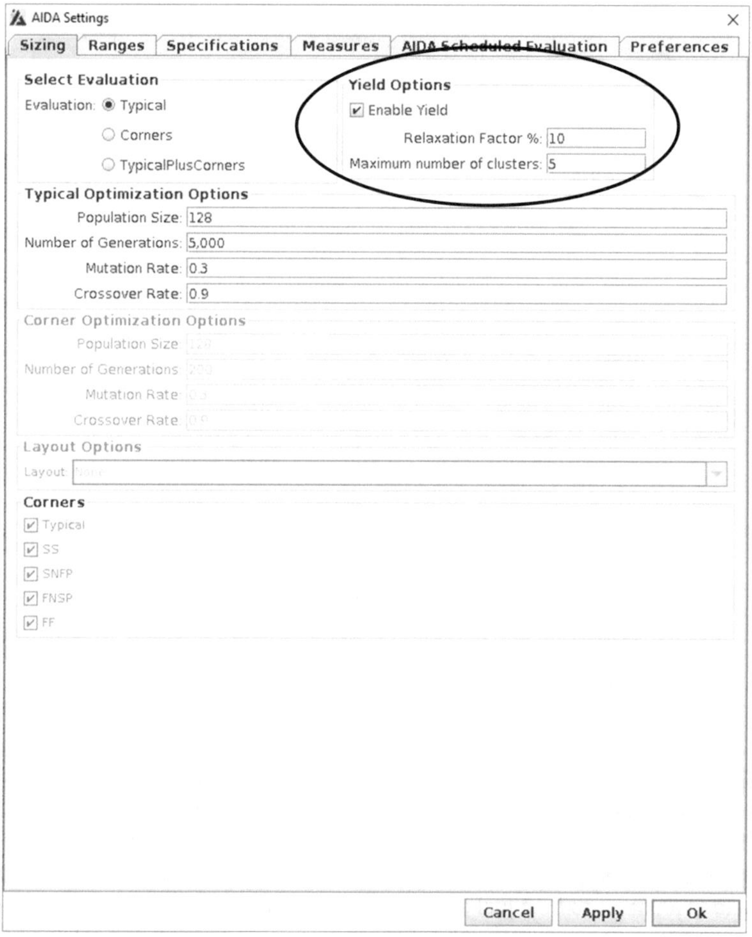

Fig. 5.21 New AIDA-C yield parameter sizing screen

saturation margins and overdrives of transistors constraints for the yield estimation process, since the solution already fulfill with these constraints for the typical conditions. Based on this description AIDA-C original GUI was changed. The modified setting screens are presented in Figs. 5.21 and 5.22.

In Fig. 5.21, the yield optimization can be enabled by the "Yield" checkbox. This screen also allows the designer to define the number of clusters and the relaxation factor. The number of clusters was included during tests and is intended for clustering algorithms where the parameter was not automatically set. The relaxation factor parameter changes the constraint/specification boundary limit value to a lower or higher value according to the inequality constraint condition, while keeping unaltered the constraint for the nominal optimization process. To illustrate the relaxation factor, consider a transistor saturation margin constraint defined as

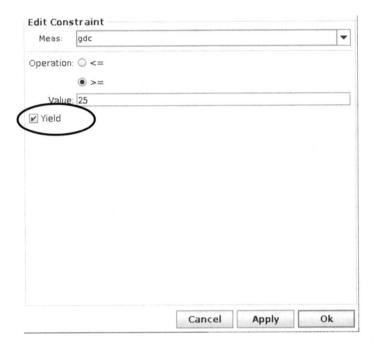

Fig. 5.22 New AIDA-C yield constraint properties screens

delta $\geq$ 70 mV using a relaxation margin of 10% the constraint is equivalent to delta $\geq$ 70 mV $-$ 10% $\rightarrow$ delta $\geq$ 63 mV, for a constraint with the reverse inequality the relaxation factor increases the constraint value, like for a consumption current constraint as $i_{DD} \leq$ 50 mA when using the same relaxation factor of 10% the constraint becomes $i_{DD} \leq$ 55 mA. In case the designer desires the conditions for the yield estimation to be equal to the typical optimization process, it is possible to set the relaxation factor to zero. The changes also made possible to enable or disable some constraints for the yield estimation process Fig. 5.22.

5.4   Conclusion

AIDA-C is a state-of-the-art analog circuit sizing tool with a user-friendly GUI. The new yield optimization methodology, now available, improves the robustness of the circuit solutions provided by the AIDA-C circuit sizing tool, particularly when smaller technology nodes are adopted for the sizing project, since new nanometer technology nodes revealed higher sensitive to parametric variability effects [13]. The implementation of the yield optimization feature was achieved by introducing the two-phase evaluation process into the already existing evaluation stage and by introducing the required modification to the GUI in order to enable the new yield optimization feature. The methodology adopted to estimate a potential circuit solution yield is based on well-tested statistical device models provided by the technology foundries, which offers three main advantages: first, the accuracy of the yield estimation, since the variability device models are provided by the foundries; second, the reduced setup time, since once the foundry PDK is installed the yield feature is available, and third, the reusability, as this yield estimation is not based on models for a specific circuit topology or some in-house developed device models, thus not requiring training the models.

References

1. N. Lourenço, R. Martins, N. Horta, *Automatic Analog IC Sizing and Optimization Constrained with PVT Corners and Layout Effects* (Springer International Publishing, Cham, 2017)
2. N. Lourenço, R. Martins, A. Canelas, R. Póvoa, N. Horta, AIDA: layout-aware analog circuit-level sizing with in-loop layout generation. Integration VLSI J. **55**, 316–329 (2016)
3. R. Martins, N. Lourenço, N. Horta, *Analog Integrated Circuit Design Automation – Placement, Routing and Parasitic Extraction Techniques* (Springer International Publishing, Cham, 2017)
4. Mentor, A Siemens Business. [Online]. Available: http://www.mentor.com. Accessed 10 Sept 2017
5. C. McAndrew, I.-S. Lim, B. Braswell, D. Garrity, Corner models: inaccurate at best, and it only gets worst. . ., in *Proc. IEEE Custom Integr. Circuits Conf.*, 2013
6. S. Bandyopadhyay, S. Saha, Some single- and multiobjective optimization techniques, in *Unsupervised Classification: Similarity Measures, Classical and Metaheuristic Approaches, and Applications*, (Springer, Berlin, 2013), pp. 17–55
7. M. Reyes-sierra, C.A.C. Coello, Multi-objective particle swarm optimizers: a survey of the state-of-the-art. Int. J. Comput. Intell. Res. **2**(3), 287–308 (2006)
8. R. Lourenço, N. Lourenço, N. Horta, *AIDA-CMK: Multi-Algorithm Optimization Kernel Applied to Analog IC Sizing* (Springer International Publishing, Cham, 2015)
9. A. Canelas, R. Martins, R. Póvoa, N. Lourenço, J. Guilherme, N. Horta, Enhancing an automatic analog IC design flow by using a technology-independent module generator, in *Performance Optimization Techniques in Analog, Mixed-Signal, and Radio-Frequency Circuit Design*, (IGI-Global, Hershey, PA, 2014), pp. 102–133
10. E. Maricau, G. Gielen, *Analog IC Reliability in Nanometer CMOS* (Springer-Verlag, New York, 2013)
11. M. Aslam, Statistical monitoring of process capability index having one sided specification under repetitive sampling using an exact distribution. IEEE Access **6**, 25270–25276 (2018)
12. Y.T. Tai, W.L. Pearn, Measuring the manufacturing yield for skewed wire bonding processes. IEEE Trans. Semicond. Manuf. **28**(3), 424–430 (2015)
13. G. Moretti, Senior Editor Chip Design Magazine, Complexity of Mixed-signal Designs, 28 Aug 2014. [Online]. Available: http://chipdesignmag.com/sld/blog/2014/08/28/complexity-of-mixed-signal-designs/. Accessed 16 Sept 2018

Chapter 6
Tests and Results

6.1 Analog IC Sizing and Optimization Tested Problems

In order to avoid repeating circuit schematics and definition of the design variables, performance specification constraints and optimization goals on the following sections of this chapter, the common circuit schematics, sizing and optimization problems adopted to test the different MC-based yield estimation and optimization methodologies are presented in this section. The design variables range and performance specification constraints boundary values presented in this section must be assumed as default values for the different circuit sizing and optimization tests presented later in this chapter. In case a set of different ranges and/or constraints are adopted on a particular test, the new optimization problem details will be presented.

The new yield estimation and optimization methodologies with a reduced time impact from the MC simulations are evaluated, mainly, by two indicators. The first indicator is the reduction rate of the number of potential solutions subject to MC simulations by each methodology when compared to performing MC simulations to all feasible potential solutions of the optimization algorithm (6.1). The tests where all feasible potential solutions were subject to MC simulations, to accurately estimate its yield value, are used as reference and from now on are identified as "Full MC."

$$R_{rate} = 1 - \#PS_{Method}/\#PS_{FullMC} \qquad (6.1)$$

where

$\#PS_{Method}$—The number of potential solutions simulated using one of the new yield estimation and optimization methodologies, which is the total number of cluster representative individuals in an optimization run.

$\#PS_{FullMC}$—The total number of feasible potential solutions simulated, which is the number of potential solutions simulated on a typical approach, where all individuals of the optimization algorithm population are simulated.

© Springer Nature Switzerland AG 2020   179
A. M. L. Canelas et al., *Yield-Aware Analog IC Design and Optimization in Nanometer-scale Technologies*, https://doi.org/10.1007/978-3-030-41536-5_6

The second indicator adopted to evaluate the new yield estimation and optimization methodologies is the accuracy of the method. The accuracy for each cluster-based methodology correspond to the yield estimation error and is computed by the root mean square error between the MC estimated yield value of each potential solutions and the estimated yield value from each of the methodologies (6.2).

$$\text{Error} = \sqrt{\frac{1}{N} \sum_{i=1}^{N} \left(\widehat{Y}_i - Y_i \right)^2} \qquad (6.2)$$

where

N—Number of feasible potential solutions tested in the full optimization process.

$\widehat{Y}_i$—Estimated yield value for potential solution i by one of the methodologies.

Y_i—Yield value for potential solution i estimated by performing MC simulations.

6.1.1   Single-Stage Amplifier with Enhanced DC Gain

The first common circuit used in the tests is a single-stage amplifier with high energy-efficiency and a gain enhancement technique [1], whose schematic is presented in Fig. 6.1.

The single-stage amplifier circuit was tested using a 130 nm silicon-based technology node and an organic top-gated carbon nanotube (OTFT) technology. The different tests adopting this circuit in the 130 nm silicon-based technology circuit considered the performance, functional and constraints specifications detailed in this section. While for the organic-based technology a set of different specifications were adopted, which are later detailed in the section where the tests results are presented. The common circuit performance specifications constraints and the devices' functional specifications constraints are presented in Table 6.1.

The circuit performance measures were calculated by electrical simulation using AC and DC analyses, considering a 6 pF capacitor as load at the output and a 3.3 V supply.

The total 18 problem design variables and respective ranges, which define the search space, are presented in Table 6.2, where the nf_ variables correspond to the transistor number of fingers, $w\_$ is the transistor channel width, and $l\_$ is the transistor length channel.

The search space is bounded by the minimum and maximum values of each design variable and discretized as a multi-dimensional grid space with uniform step sizes for each dimension, defined by the column "Grid Step" in Table 6.2.

The optimization problem objectives for the single-stage amplifier with enhanced DC gain of Fig. 6.1 are the maximization of the parametric yield and the maximization of the circuit's figure-of-merit (FoM) for energy-efficiency:

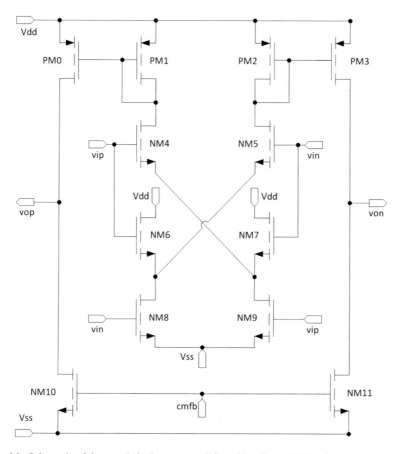

Fig. 6.1 Schematic of the tested single-stage amplifier with enhanced DC gain

Table 6.1 Single-stage amplifier performance and functional specification constraints for the silicon-based technology node

Performance specifications		
I_{DD} (μA)	Current consumption	≤ 350
Gain DC (dB)	Low-frequency gain	≥ 50
GBW (MHz) @ $C_{load} = 6$ pF	Gain-bandwidth product	≥ 30
PM (°)	Phase margin	≥ 60
FoM (MHz × pF/mA)	Figure of merit (6.3)	≥ 850
Functional specifications	$V_{DS} - V_{DSat}$ (mV)	$V_{GS} - V_{TH}$ (mV)
PMOS	≥ 70	≥ 100
NMOS	≥ 70	≥ 50

Table 6.2 Single-stage amplifier optimization problem design variables and ranges for the silicon-based technology node

Variable	Unit	Min	Grid step	Max
nf0, nf1, nf4, nf6, nf8, nf10[a]	–	1.0	2.0	8.0
14, 16, 18, 110[b]	μm	0.34	0.01	0.94
w0, w1, w4, w6, w8, w10[c]	μm	1.0	0.10	100
10, 11[d]	μm	0.30	0.01	0.90

[a]Number of Gate Fingers for the devices pair PM0–PM3, PM1–PM2, NM4–NM5, NM6–NM7, NM8–NM9, and NM10–NM11, correspondingly
[b]Gate Finger Length for the devices pair NM4–NM5, NM6–NM7, NM8–NM9, and NM10–NM11, correspondingly
[c]Gate Finger Width for the devices pair PM0–PM3, PM1–PM2, NM4–NM5, NM6–NM7, NM8–NM9, and NM10–NM11, correspondingly
[d]Gate Finger Length for the devices pair PM0–PM3 and PM1–PM2, correspondingly

$$\text{FoM} = \frac{\text{GBW} \times C_{\text{load}}}{I_{\text{DD}}} \left[\frac{\text{MHz} \times \text{pF}}{\text{mA}} \right] \tag{6.3}$$

where

GBW—Gain-bandwidth product
C_{load}—Load capacity
I_{DD}—Current consumption

For the silicon-based technology, the foundry PDK statistical device models to perform the MC simulations, considering process and mismatch variations, were used. The 130 nm silicon-based technology node models variation parameters as Gaussian distributions. For transistor devices a total of 14 parameters were considered, where 2 parameters are for device mismatch and 12 for process variation. Among those parameters are the gate oxide thickness, bottom junction capacitance, threshold voltage, saturation velocity, low-field surface mobility, and width and length fitting parameter. In Table 6.3, the model parameters from BSIM3V3 MOSFET model affected by variability in the selected 130 nm technology node are listed.

6.1.2   Grounded Active Inductor

The second circuit used in the tests is a grounded active inductor with an improved topology based on Manetakis regulated cascode active inductor comprising three control voltages for tunability [2]. The circuit was design for a 130 nm silicon-based technology and in Fig. 6.2 the circuit schematics is presented. The optimization goals for this circuit are the maximization of both objectives, the parametric yield and the inductor quality factor. In Table 6.4, the optimization problem constraints are detailed and in Table 6.5 the design variables are presented. The supply was 1.8 V and all transistors adopt the technology node smallest gate length, $l_{\text{min}} = 120$ nm,

Table 6.3 BSIM3V3 MOSFET model parameters affected by variability in the adopted 130 nm technology node

Device	Variability	Parameter	Description
Transistor	Process	tox	Gate-oxide thickness.
		cj	Bottom junction capacitance per unit area.
		cjsw	Source/drain sidewall junction capacitance per unit length.
		cjswg	Source/drain gate sidewall junction capacitance per unit length.
		vth0	Threshold voltage
		ags	Gate bias coefficient of the Abulk
		lvth0	Length dependence of vth0 binning parameter
		lint	Length offset fitting parameter from I–V without bias
		vsat	Saturation velocity
		wvth0	Width dependence of vth0 binning parameter
		wint	Width offset fitting parameter from I–V without bias
		pvth0	Product dependence of vth0 binning parameter
	Mismatch	vth0	Threshold voltage
		u0	Low-field surface mobility at t_{nom}

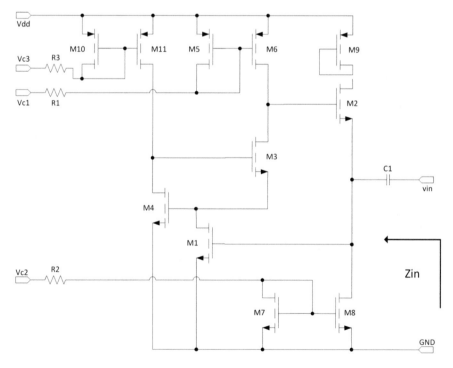

Fig. 6.2 Active inductor schematics

Table 6.4 Performance and functional specification constraints for the active inductor circuit

Performance specifications		
$f_{res\ max}$ (GHz)	Maximum operating frequency	≥ 16
BW (GHz)	Bandwidth	≥ 4
C (fF)	Parasitic capacitance at operating frequency	≤ 50
$L = Im(Z_{in})/w$ at f_{Qmax} (pH)[a]	Inductance at operating frequency	≥ 200
$Q = Im(Z_{in})/Re(Z_{in})$ $(-)$[a]	Inductor quality factor at operating frequency	≥ 100
Functional specifications	$V_{DS} - V_{DSat}$ (mV)	$V_{GS} - V_{TH}$ (mV)
PMOS	≥ 70	≥ 100
NMOS except M4	≥ 70	≥ 100
Transistor M4	≥ 70	≥ 20

[a]$Im(x)$ and $Re(x)$ return the imaginary and real part of x respectively

Table 6.5 Optimization problem design variables and ranges for the active inductor circuit

Variable	Unit	Min	Grid step	Max
wfpmos[a], wfnmos[b]	μm	1.5	0.1	7.2
M_M1, M_M2, M_M3, M_M4[c]	–	1	2	10
Wcin[d]	μm	10	0.1	100
wf1, wf2, wf3, wf4[e]	μm	1.5	0.1	7.2
V_{C1}, V_{C2}, V_{C3}[f]	V	0.3	0.1	1.8

[a]Gate Finger Width for the devices M5, M6
[b]Gate Finger Width for the devices M7, M8
[c]Multiplier for devices M1, M2, M3 and M4, correspondingly
[d]Width and Length of capacitor C1
[e]Gate Finger Width for the devices M1, M2, M3 and M4, correspondingly
[f]Bias voltages

and all the resistors have the same fixed values of width and length, $w_R = 2$ μm and $l_R = 5$ μm, respectively.

The active inductor circuit devices belong to the PDK radio frequency (RF) CMOS library, modelling *regular* transistors M5 to M11 variation effects by a total of 20 parameters for each device, where 18 parameters are for modelling process variations and 2 for mismatch variations. Transistors M1 to M4 are triple-well devices, adding four more process parameters to the previous *regular* transistors' variation parameters. The variation effects in resistors are modelled by a total of seven parameters, where six parameters are for modelling process variations and one for mismatch variations. Capacitors models from the RFCMOS variability library use a total of five parameters to simulate variations, having four parameters for process variations and one for mismatch.

6.1.3 Low-Noise Amplifier for 5 GHz Applications

The next circuit, Fig. 6.3, is a two-stage low-noise amplifier (LNA) for 5 GHz applications [3]. The circuit was designed using a 130 nm silicon-based technology

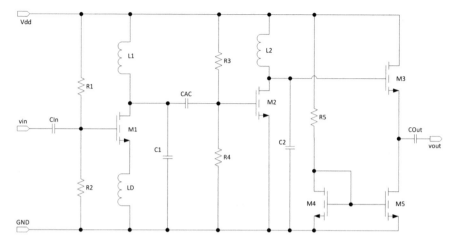

Fig. 6.3 Low-noise amplifier for 5 GHz applications schematics

Table 6.6 LNA performance and functional specification constraints

Performance specifications		
I_{DD} (A)	Current consumption of the LNA	$\leq 8e-3$
S_{11} (dB)	Input reflection coefficient @ 5.05 GHz	≤ -14
S_{21} (dB)	Forward gain @ 5.05 GHz	≥ 19
S_{12} (dB)	Reverse gain @ 5.05 GHz	≤ -40
S_{22} (dB)	Output reflection coefficient @ 5.05 GHz	≤ -14
NF (dB)	Noise figure @ 5.05 GHz	≤ 2.4
BW (MHz)	-3 dB Bandwidth at S_{21} maximum peak	≥ 700
ZinRE (Ω)	Min. real part of input impedance (1–100e9 Hz)	>0
ZoutRE (Ω)	Min. real part of output impedance (1–100e9 Hz)	>0
P1dB (dBm)	Input referred 1 dB gain compression point	≥ -29
Functional specifications	*NMOS*	*M5*
$V_{DS} - V_{DSat}$ (mV)	≥ 100	
$V_{GS} - V_{TH}$ (mV)	≥ 80	
VDS (mV)		≥ 450 and ≤ 550

node and a supply of 1.2 V. The circuit performance specifications were estimated by electrical simulations using DC, AC, noise, and steady-state analyses.

The performance and functional specification constraints are listed in Table 6.6, and the design variables and respective ranges are detailed in Table 6.7. The nf_ design variables correspond to the transistor number of fingers, w_ is the transistor channel width, and l_ is the transistor length channel. The wc_ design variables are the width and length of the capacitors. While for inductors, the design variables are od_ as the outer diameter and wl_ as the metal width. Three optimization goals were

Table 6.7 LNA optimization problem design variables and ranges

Variable	Unit	Min	Grid step	Max
wcin, wc1, wcac, wc2, wcout[a]	µm	10.0	0.1	100.0
nf1, nf2, nf3, nf5[b]	–	4.0	1.0	16.0
l1, l2, l3[c]	µm	0.12	0.01	0.36
w1, w2, w3, w5[d]	µm	0.9	0.1	7.2
odld, odl1, odl2[e]	µm	118.0	0.01	299.0
wld, wl1, wl2[f]	µm	2.2	0.2	9.6

[a]Width of the symmetrical capacitors CIN, C1, CAC, C2, and COUT correspondingly
[b]Number of Gate Fingers of the devices M1, M2, M3, and M5 correspondingly
[c]Gate Finger Length of the devices M1, M2, M3, and M5 correspondingly
[d]Gate Finger Width of the devices M1, M2, M3, and M5 correspondingly
[e]Outer Diameter of the inductors LD, L1, and L2 correspondingly
[f]Metal Width of inductors LD, L1, and L2 correspondingly

set, having as objectives the maximization of the forward gain (S_{21}) and yield, and the minimization of the current consumption (I_{DD}) drained from the voltage supply source.

The adopted foundry PDK statistical device models for the 130 nm technology node, as in the previous examples, models parameters variations as Gaussian distributions offering to IC designers the ability of setting the range of parameters variations in multiples of the σ of the distribution. In the LNA circuit example and in all the other test circuits where the 130 nm technology node was selected, the range for the parameter's variation was set to 3σ. Thus, every variability parameter in the statistical device model will assume values in the range $[3\sigma - \mu, \mu + 3\sigma]$, where μ is the parameter distribution mean value and σ is the standard deviation.

The variability models for capacitors use five statistical parameters to model variation, being one of them for mismatch variations. The resistors' variation models adopt a total of seven parameters, one of those parameters is intended to model mismatch variations. Transistors' variations are modelled by a total of 14 parameters, where 2 are for device mismatch and 12 for process variation. The process variation in inductors is modelled by selecting one of the three available libraries, where in each library the inductors quality factor assumes a different value.

6.1.4 Sub-µW Tow-Thomas-Based Biquad Filter with Improved Gain

The last circuit example implements a biquad filter based on a second-order Tow-Thomas architecture, Fig. 6.4. The proposed implementation is a sub-µW design, specifically developed for biomedical and healthcare circuits and systems

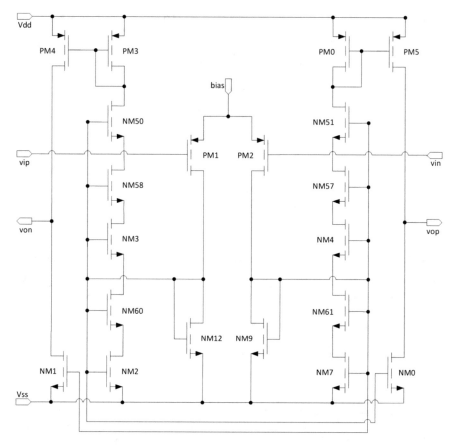

Fig. 6.4 Sub-μW Tow-Thomas-based biquad filter with improved gain schematics

(waiting for publication). Originally the circuit was designed in a 130 nm technology node but to assess the impact of parametric variations in smaller technology nodes, the example presented in this work was designed in a 65 nm technology node using a 0.9 V supply.

The circuit performance and functional specifications constraints are listed in Tables 6.8 and 6.9 and 31 optimization design variables are detailed. To maximize the filter gain, PM1 and PM2 transistor are working in the subthreshold region, which is imposed by the defined functional constraints.

The PDK variability model files for the adopted 65 nm technology node are encrypted, making impossible detailing the number of variability parameters modelled and its probability distributions. However, the variation range definition for those parameters is similar to the 130 nm technology and was set $\pm 3\sigma$.

Table 6.8 Tow-Thomas-based biquad filter performance and functional specification constraints

Performance specifications		
I_{DD} (μA)	Current consumption	≤ 0.5
Gain DC (dB)	Low-frequency gain	≥ 25
GBW (kHz)	Unity gain frequency	≥ 100
PM (°)	Phase margin	≥ 50
Sdnoise (mV/$\sqrt{\text{Hz}}$)	Output noise spectral density	≤ 0.5
Offset (mV)	Open loop offset	≤ 50
BW (Hz)	-3 dB Bandwidth	≥ 100
Functional specifications	$V_{DS} - V_{DSat}$ (mV)	$V_{GS} - V_{TH}$ (mV)
NMOS		≤ -100
PMOS except PM1 and PM2		≤ -100
PM1 and PM2	≥ 200	≥ -0.1

Table 6.9 Tow-Thomas-based biquad filter optimization problem design variables and ranges

Variable	Unit	Min	Grid step	Max
wpm5, wpm3, wpm1, wnm60, wnm56, wnm50, wnm3, wnm2, wnm12, wnm0[a]	μm	0.5	0.01	8.0
nfpm5, nfpm3, nfpm1, nfnm60, nfnm56, nfnm50, nfnm3, nfnm2, nfnm12, nfnm0[b]	–	2.0	2.0	32.0
lpm5, lpm3, lpm1, lnm60, lnm56, lnm50, lnm3, lnm2, lnm12, lnm0[c]	nm	60.0	5.0	360.0
ib[d]	nA	50.0	10.0	500.0

[a]Gate Finger Width of the devices pair PM5–PM0, PM3–PM4, PM1–PM2, NM60–NM61, NM56–NM57, NM50–NM51, NM3–NM4, NM2–NM7, NM12–NM9, NM0–NM1 correspondingly
[b]Number of Gate Fingers of the devices pair PM5–PM0, PM3–PM4, PM1–PM2, NM60–NM61, NM56–NM57, NM50–NM51, NM3–NM4, NM2–NM7, NM12–NM9, NM0–NM1 correspondingly
[c]Gate Finger Length of the devices pair PM5–PM0, PM3–PM4, PM1–PM2, NM60–NM61, NM56–NM57, NM50–NM51, NM3–NM4, NM2–NM7, NM12–NM9, NM0–NM1 correspondingly
[d]Bias current

6.2 Comparison Between KMS, FUZYE, and Hierarchical Agglomerative Clustering

This section aims to identify among the three clustering algorithms, i.e., KMS, FUZYE using the FCM algorithm, and Hierarchical Agglomerative Clustering (HAC), which provides the best results in terms of the number of solutions subject to MC simulations and accuracy. Additionally, from the comparison among the different clustering techniques results, only two best clustering techniques will be adopted for the yield optimization tests in the rest of the chapter.

Two circuits were adopted for the clustering comparison tests: the silicon-based 130 nm technology node single-stage amplifier with enhanced gain, Sect. 6.1.1, and the 130 nm technology node grounded active inductor, Sect. 6.1.2. For each circuit five runs for each methodology were performed.

The comparison tests adopt for the GA parameters a population of 256 individuals and as stopping criterion a maximum of 300 generations. For each potential solution subject to MC simulations 1000 iterations by the Eldo™ electrical simulator using the foundry PDK's statistical device models, considering process and mismatch variations, were performed.

The clustering algorithms parameters were set according to:

– The KMS-based methodology set to five the number of clusters for both circuits.
– The FUZYE methodology, with the fuzziness parameter set to $m = 2$, will automatically adjust the number of clusters at each generation of the optimization algorithm.
– The HAC algorithm stopping level line parameter was set at a distance of one, as described in Sect. 4.3.5. Being the distance computed as the Euclidian distance between data points. By setting the stopping line level to a fixed value, at each generation a different number of clusters are applied depending on the data distribution. The number of clusters is important, since one individual is selected from each cluster to perform MC simulations, and a higher number of clusters represents a larger number of MC simulations.

During each of the runs, the number of potential solutions subject to MC simulations by each of the methodologies was registered and compared to performing a Full MC optimization process to assess the reduction rate of each methodology (R_{rate} (6.1)), also the yield estimation error was computed according to (6.2).

According to the results reported in Table 6.10, the best methodology in terms of accuracy is the HAC but the gains in accuracy with respect to the other two methodologies are obtained by performing a higher number of MC simulations.

The HAC performs on average almost two times more MC simulations than the FUZYE methodology, which is the best in terms of simulations reduction rate. Adjusting the stopping line level, in the HAC-based methodology, to reduce the number of simulated solutions in order to achieve reduction rates above 85%, causes an increase in the yield estimation error to 3.65, above the error of the other two

Table 6.10 Comparison for the KMS, FUZYE, and HAC methodologies in terms of potential solutions subject to MC simulations and accuracy for the single-stage amplifier

	Full MC	KMS-based			FUZYE			HAC-based		
			R_{rate}			R_{rate}			R_{rate}	
Run	Simul.	Simul.	(%)	Error	Simul.	(%)	Error	Simul.	(%)	Error
1	6687	1252	81.28	3.40	957	85.69	3.08	2123	68.25	2.03
2	9482	1238	86.94	4.25	958	89.90	3.79	1640	82.70	3.83
3	7632	1129	85.21	2.37	808	89.41	2.07	1454	80.95	2.11
4	9831	1307	86.71	3.91	1017	89.66	3.47	2030	79.35	3.25
5	8548	1264	85.21	3.56	1055	87.66	2.83	2007	76.52	2.75
Average	8436	1238	85.32	3.50	959	88.63	3.05	1851	78.06	2.79

Single-stage amplifier with enhanced DC gain

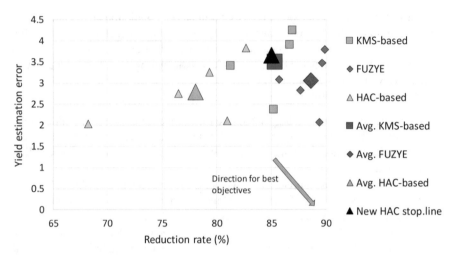

Fig. 6.5 Graphical representation of KMS, FUZYE, and HAC methodologies reduction rate and yield estimation error for the single-stage amplifier

Table 6.11 Comparison for the KMS, FUZYE, and HAC methodologies in terms of potential solutions subject to MC simulations and accuracy for the grounded active inductor

Grounded active inductor										
	Full MC	KMS-based			FUZYE			HAC-based		
Run	Simul.	Simul.	R_{rate} (%)	Error	Simul.	R_{rate} (%)	Error	Simul.	R_{rate} (%)	Error
1	4572	1396	69.47	3.13	1376	69.90	2.58	1574	65.57	3.35
2	4820	1369	71.60	3.65	1356	71.87	2.94	1551	67.82	3.59
3	5099	1461	71.35	3.00	1058	79.25	2.69	1820	64.31	2.39
4	3643	1461	59.90	2.74	1460	59.92	2.21	1530	58.00	2.28
5	5925	1408	76.24	3.68	810	86.33	3.56	1382	76.68	3.50
Average	*4812*	*1419*	*70.51*	*3.24*	*1212*	*74.81*	*2.80*	*1571*	*67.34*	*3.02*

tested methodologies. In Fig. 6.5, a graphical representation for the reduction rate and yield estimation error of the different tested methodology is depicted. The graphic shows that the average FUZYE result dominates the average KMS-based methodology and if the HAC run with new stopping line value is considered, then FUZYE also dominates the HAC-based with improved reduction rate method.

In Table 6.11, the grounded active inductor results are reported. This circuit is much more sensitive to variability effects resulting in smaller number of feasible solutions at each generation. This fact causes a smaller simulations reduction rate for all the methodologies. The best methodology for both indicators, i.e., reduction rate and error, is the FUZYE methodology. In this test circuit was also manually improved, by adjusting the stopping line level, the simulations reduction rate of the HAC-based methodology, with results similar to the previous circuit tests, i.e., an increase in the reduction rate leads to an increase in the yield estimation error, which resulted in a dominance from FUZYE with respect to the other methodologies.

Considering the simulations reduction rate, the best two methodologies are the FUZYE and the KMS-based method. Observing the yield estimation error reveals that the best two methodologies are the FUZYE and the HAC-based method. Overall, based on the tests results and the adopted indicators, the FUZYE methodology is the best among all the tested methodologies. Although the HAC-based methodology is more accurate than KMS-based methodology, the reality is that the time impact of a larger number of MC simulations may discourage the adoption of an IC sizing and optimization approach based on the HAC. Additionally, when the HAC-based methodology simulation rate was improved by adjusting the stopping line level, the new yield estimation error exceeded the error of the KMS-based methodology. Based on these conclusions, the HAC-based methodology was discarded and for the rest of the tests only FUZYE and the KMS-based methodology will be adopted.

6.3 Clustering Algorithms Runtime and Memory Usage

The new yield estimation and optimization methodology based on MC simulations main concern is having the smallest time impact into the already implemented IC sizing and optimization processes of AIDA. Another concern is the memory requirements, since population-based optimization algorithms often require large populations to solve optimization problems.

In order to identify which stages of the new yield estimation and optimization methodology may require larger amounts of memory and have higher executions times, the new java implemented module for the yield estimation was divided into the following four sections:

1st Section—Identifies feasible potential solutions and assigns negative yield value to infeasible potential solutions.
2nd Section—Cluster potential solutions and identify the cluster representative individuals.
3rd Section—Perform MC simulations on the cluster representative individuals.
4th Section—Estimate the yield value for all feasible potential solutions.

The runtime of each section was registered for the KMS-based and FUZYE methodologies during a sizing and optimization process with a stopping criterion of 300 generations and different population sizes of 256, 512 and 1024 individuals. The circuit adopted in this test was the single-stage amplifier circuit of Sect. 6.1.1, and 500 iterations/MC simulation on each cluster representative individual were performed by the Eldo™ electrical simulator. Also, the total memory requirements were registered for each methodology, considering the heap memory and non-heap memory usage computed by the following command in Fig. 6.6.

```
long getTotalMemoryUsage() {

  return ManagementFactory.getMemoryMXBean().getHeapMemoryUsage().getCommitted() +

  ManagementFactory.getMemoryMXBean().getNonHeapMemoryUsage().getCommitted();

}
```

Fig. 6.6 Code to compute the total memory usage

Table 6.12 Average memory usage for both KMS-based and FUZYE methodologies

	KMS-based		FUZYE	
Sections	No garbage col. (MB)	With garbage col. (MB)	No garbage col. (MB)	With garbage col. (MB)
1st	72.38	69.15	157.50	74.92
2nd	72.50	69.15	158.06	74.93
3rd	106.52	69.40	160.63	75.22
4th	72.62	69.13	159.58	74.93
Highest values	*106.52*	*69.40*	*160.63*	*75.22*

For registering the memory usage two different runs were performed. The first run measures the memory usage at different phases inside each section, registering the highest value of memory usage. The second run, before each section performs several garbage collection operations until the value of the memory in use settles, this memory cleaning operations assures that all java objects no longer in use are removed from memory. After the garbage collection operation, the memory usage of each section is registered as in the first memory usage run. Like in the runtime tests, the memory usage run with garbage collection was repeated using an increased number of individuals of the optimization algorithm population. The tests were carried out on PC with an Intel© i7-3770 CPU with 16 GB of memory run and for the KMS-based methodology the number of clusters was set to five.

In Table 6.12, the average memory usage results are presented. The values reported in Table 6.12 are the average of the maximum memory usage per section for the all the generations of the optimization process where feasible potential solutions exist. As expected, the highest memory usage occurs when the electrical simulator is invoked, and the results are collected from the simulator output files. The impact of forcing the garbage collection results in an average saving of 12% and 52% memory usage for the KMS-based and FUZYE methodologies, respectively. The memory usage results revealed that the new yield estimation module does not require large amounts of memory for its execution.

In Table 6.13, the results for an increased number of individuals in the optimization algorithm are presented. As expected, when the optimization algorithm adopts a larger population, the memory usage also increases. Memory usage has its

Table 6.13   Average memory usage for both KMS-based and FUZYE methodologies with different optimization algorithm population sizes

	KMS-based memory usage (MB)			FUZYE memory usage (MB)		
Sections	256 Indiv.	512 Indiv.	1024 Indiv.	256 Indiv.	512 Indiv.	1024 Indiv.
1st	69.15	82.67	124.58	74.92	81.64	126.79
2nd	69.15	82.84	123.51	74.93	81.98	125.24
3rd	69.40	116.19	157.02	75.22	99.72	143.61
4th	69.13	82.60	124.61	74.93	81.69	126.86
Highest values	*69.40*	*116.19*	*157.02*	*75.22*	*99.72*	*143.61*

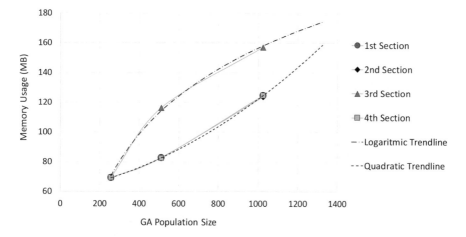

Fig. 6.7   KMS memory usage vs. population size

maximum value at the section where the electrical simulator is invoked, and simulation results are collected to estimate the yield of the tested potential solutions.

In Fig. 6.7, using the values from Table 6.13 was plotted a graphic showing the memory usage for the KMS-based methodology. Additionally, trendlines were added to the graphic, to show the relation between memory usage and population size. This graphic shows that with exception of the third section, where the simulator is invoked, the memory usage at each section progresses very similarly with the increase of population. Considering the trendlines that better adjust to the data points shows that the 3rd section memory usage has logarithmic behavior with the increase of the number of individuals in the population. Whereas the rest of the sections have a quadratic behavior, given by the trendline quadratic equation (6.4).

$$\text{Trend}_{\text{1st,2nd and 4th Sections}} = 0.0003781 \text{ Indiv.}^2 + 0.0238 \text{ Indiv.} + 69.15 \quad (6.4)$$

where Indiv. is the number of individuals in the optimization algorithm population.

The FUZYE memory usage for different population sizes is depicted in Fig. 6.8. The graphic shows that the 3rd sections memory usage grows in a linear manner with

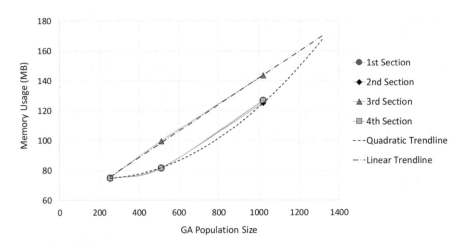

Fig. 6.8 FUZYE memory usage vs. population size

respect to the increase of the population, whereas the other sections follow a quadratic growth like in the KMS-based methodology.

In Table 6.14, the average runtime for each section per generation of the yield estimation module, considering the different population sizes tested, are reported. The presented runtime values, as in the memory usage tests, are the sections average runtime values computed for all the generations of the optimization process where feasible potential solutions exist. In the last row, the total yield estimation module runtime is presented, i.e., the sum of all section's runtime for all the optimization process. The column with the header "% Total runtime" represents the average percentage of total runtime of each section with respect to the total yield estimation module runtime.

Table 6.14 total runtime results show that FUZYE is faster than the KMS-based methodology. The better runtime performance is related to the smaller number of potential solutions simulated at each generation of the optimization algorithm by the FUZYE methodology. The 2nd section runtime refers to the average runtime that each methodology spends performing the clustering task and identifying the cluster representative individual. The average clustering task is concluded in 1.5 ms by the KMS algorithm, while FCM algorithm takes on average 240 ms. The clustering runtime difference between both methodologies is caused by the automatic cluster number setting technique implemented in the FUZYE methodology, which performs additional clustering iterations to determine the correct cluster number.

As expected, the largest amount of the total runtime is consumed by an external process to the yield estimation module, the electrical simulator. Tests launching manually the electrical simulator, using this section MC netlist and test bench files with five potential solutions to perform MC simulations, showed that on average 18s were spent to complete the simulations. Results also show that for the KMS-based methodology the simulation time is similar for all population sizes; this is explained by the fact the number of clusters is constant, and thus the same amount of MC

Table 6.14 Average runtime for both KMS-based and FUZYE methodologies

Sections	KMS-based					FUZYE				
	Avg. runtime (ms)			% Total		Avg. runtime (ms)			% Total	
	256 Indiv.	512 Indiv.	1024 Indiv.	Runtime		256 Indiv.	512 Indiv.	1024 Indiv.	Runtime	
1st	1.50	1.79	1.87	0.01		1.04	1.26	3.24	0.01	
2nd	1.01	0.93	2.01	0.01		85.83	92.60	541.65	1.65	
3rd	20,951.43	21,056.00	21,010.75	96.50		12,198.07	13,823.52	13,857.87	94.79	
4th	723.68	757.49	822.68	3.48		432.23	528.47	537.91	3.55	
Total runtime (hh:mm)	01:35	01:39	01:40			00:57	01:06	01:10		

simulations is performed for each of the tested population sizes. A similar situation occurs for the FUZYE methodology; although the number of clusters is potentially different at each generation, the fact is that the maximum number of clusters is limited by the square root of the number of feasible potential solutions at each generation, which is almost the same when large population sizes are adopted by the tests. In conclusion, the simulations runtime is the bottleneck of the yield estimation module, which is a factor outside our control. The simulation process can only be speedup by adopting parallelization techniques, which are now offered by most of the commercial electrical simulators. The major problem of adopting parallelization is that it requires a larger number of software licenses, one per parallel process launched, and most IC design companies have a limited number of licenses for the electrical simulator.

6.4 Yield Optimization Using KMS-Based Clustering Algorithm Results

The technique described in Sect. 4.3.2, using the KMS algorithm for clustering and selecting the cluster representative individual as the potential solution with best objective values [4], was tested in two different technologies. For the first test a 130 nm silicon-based technology node was used with single-stage amplifier with enhanced DC gain, Sect. 6.1.1. The second test implements the same circuit in an organic top-gated carbon nanotube OTFT technology.

The initial tests using the modified and hybrid KMS-based methodology are intended to understand the behavior and potential of the new yield estimation and optimization methodology inside the AIDA-C optimization loop.

6.4.1 Silicon Technology Single-Stage Amplifier with Enhanced DC Gain

The sizing and optimization process for the silicon-based technology single-stage amplifier use the functional and performance specification constraints detailed in Table 6.1 and the design variables presented in Table 6.2. In addition to the default constraints from Table 6.1 was set the constraint "yield $\geq$ 60%." The optimization process goals were the maximization of the yield and the circuit FoM given by (6.3). The NSGA-II optimization kernel was set to run for 300 generation with a population of 256 individuals. Additionally, a MC netlist and test bench file was prepared; in this file 500 MC iterations were defined and the foundry PDK's statistical device models were invoked, considering both process and mismatch variations. The number of MC iterations per simulation was computed by (6.5) [5, 6], which is similar to (3.8), where the yield (Y) parameter was set to 90%, with a confidence

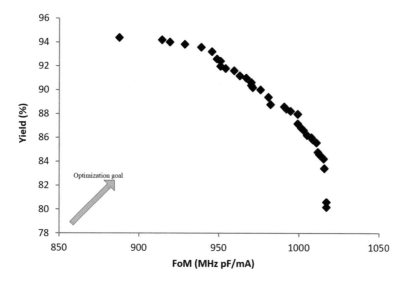

Fig. 6.9 Final Pareto front for the yield optimization process

level of 99.7% corresponding to 3σ, thus $C_\sigma = 3$, and with a confidence interval of $\varepsilon = 4\%$.

$$N = \left(\frac{C_\sigma}{\varepsilon}\right)^2 Y(1 - Y) \qquad (6.5)$$

The number of clusters was set to ten for the entire optimization process, and the tests were carried out on a PC with an i7-3770 Intel© CPU with 16 GB of RAM.

The optimization process, with the previously defined parameters, took almost 80 min for completion and reached the set of solutions presented in the Pareto optimal front shown in Fig. 6.9. The obtained set of solutions reach as far as 94.4% yield with a FoM = 888 MHz pF/mA to a yield of 80.2% for a FoM = 1017 MHz pF/mA using as relaxation factor 0%, Fig. 5.21, where all performance specifications constraints were considered to estimate the yield. In the current test scenario, it was also recorded a reduction in the total number of MC simulations of 75% when comparing to a Full MC optimization process where all feasible potential solutions are simulated.

In order to assess the impact of the relaxation parameter, new tests setting to 5% and 10% the relaxation in the performance specifications constraints were performed and compared to the initial test with 0% relaxation factor, Fig. 6.10.

The Pareto fronts presented in Fig. 6.10 show that by increasing the relaxation factor it was possible to achieve solution with higher FoM and, naturally, higher yield values. This result is explained by the fact that the relaxation factor expands the acceptability region for potential solutions in terms of yield. Notice that the feasibility region is defined by the constraints defined for the typical specifications, which are evaluated at the first evaluation stage. The effect of increasing the relaxation factor parameter resulting in higher yield values is depicted in Fig. 6.11. Where a

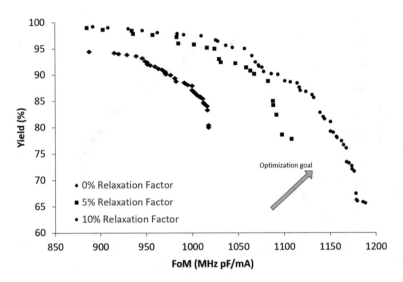

Fig. 6.10 Sizing and optimization runs with different values of relaxation factor in the performance specifications

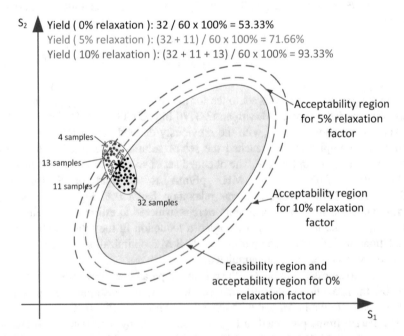

Fig. 6.11 Impact of increasing the relaxation factor in the estimated yield value

potential solution near the boundary of the feasibility region has its yield value increased since more MC samples become accepted as valid in terms of the new relaxed performance specifications constraints caused by the increase in the relaxation factor.

6.4.2   Organic Process Design Kit Single-Stage Amplifier with Enhanced DC Gain

Considering the fabrication of the single-stage amplifier circuit using inkjet technology, the production variability effects were modelled in the circuit netlist by taking into account the printer specifications and recent works [7] performed in a Fujifilm Dimatix DMP-2831inkjet printer.

In order to perform MC simulations, the standard silicon-based PDKs provide statistical variability device models. The lack of this type of models in the adopted Organic Process Design Kit (OPDK) represents an evident difficulty in performing the MC-based yield optimization process. To overcome this issue, the circuit netlist was modified to include the process variability introduced by the Dimatix printer. The circuit netlist incorporates the reported drop size repeatability error into the W and L design variables of the transistors [8]. The second type of error introduced by the printing process, drop placement accuracy, was not considered because this error is only caused when the printer cartridge is replaced, and with a carefully planned printing strategy it is possible to print each circuit layer with only one cartridge.

The drop size repeatability printing error was modelled as Gaussian distribution centered in the transistor nominal W and L, with standard deviations of 0.5%.

The yield optimization process performed 500 MC iterations per generation and cluster, with a 99.7% confidence level. The number of clusters was defined by using the Elbow method and was set to 6. The performance and functional specifications are detailed in Table 6.15. Also, the design variables are presented in Table 6.16.

Table 6.15 Performance and functional specification constraints

Performance specifications				
I_{DD} (mA)	Gain (dB)	GBW (kHz) @ $C_{load} = 37$ pF	PM (°)	FoM (kHz × pF/mA)
≤ 9	≥ 26	≥ 1	≥ 47	≥ 100
Functional specifications			$V_{DS} - V_{DSat}$ (V)	$V_{GS} - V_{TH}$ (V)
N-OTFT			≥ 0.2	≥ 0.2
P-OTFT			≥ 0.1	≥ 0.1

Table 6.16 Optimization problem variables and ranges

Variable	Min (μm)	Grid unit (μm)	Max (μm)
l0, l1, l4, l6, l8, l10[a]	40	1	200
w0, w1, w4, w6, w8, w10[b]	50	1	1000

[a] Gate Finger Width for the devices pair PM0–PM3, PM1–PM2, NM4–NM5, NM6–NM7, NM8–NM9, and NM10–NM11, correspondingly
[b] Gate Finger Length for the devices pair PM0–PM3, PM1–PM2, NM4–NM5, NM6–NM7, NM8–NM9, and NM10–NM11, correspondingly

The optimization objectives were the maximization of the Yield, GDC, and FoM. The FoM in this example is defined by (6.6). This circuit was biased at 20 V.

$$\text{FoM} = \frac{\text{GBW} \times C_{\text{load}}}{I_{\text{DD}}} \left[\frac{\text{kHz} \times \text{pF}}{\text{mA}}\right] \qquad (6.6)$$

The optimization process was set to run for 1000 generations with a population of 256 individuals, which took almost 5 h for completion. The yield-aware optimization solutions considering the objectives Yield, GDC, and FoM are presented in Fig. 6.12.

In Table 6.17 solutions that correspond to the best optimization values per objective are presented, also the tested solution identified by a circle in Fig. 6.12 is detailed in the last gray line of table.

The obtained solutions achieve maximum yield values up to 72.3%; in the FoM objective it was possible to obtain values of 1078 kHz × pF/mA, and on the other extreme of the objective Pareto surface, a solution with a DC gain of 32.7 dB was achieved. For proof-of-concept, an AC and two transient analyses were made using Mentor's Eldo[®], for an intermediate solution, circled in Fig. 6.12. The results are shown in Fig. 6.13. It is observable that the tested solution is valid, since the frequency-domain and the time-domain responses are commonly shaped waveforms, and according to the expected measurements in the optimization process.

The selected solution was also tested in 3000 MC iterations. The MC run, depicted in Fig. 6.14, reveals a final yield of 64.5% with precise results.

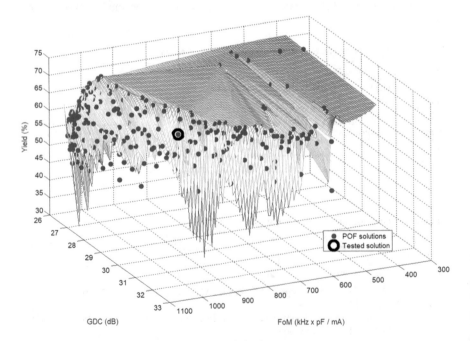

Fig. 6.12 Pareto surface for the yield-aware optimization process

Table 6.17 OPDK yield-aware best solutions

Yield	FoM	GBW	GDC	IDD	PM
[%]	[kHz×pF/mA]	[kHz]	[dB]	[mA]	[°]
72.3	465.6	111.5	27.4	8.86	49.5
51.3	1078.2	125.0	26.8	4.29	47.0
67.3	588.0	120.2	32.7	7.56	47.0
65.0	919.7	116.2	30.1	4.67	47.2

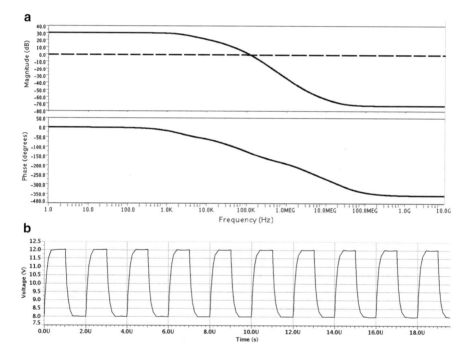

Fig. 6.13 OPDK single-stage amplifier (**a**) AC response. (**b**) Step response in a unitary gain configuration

6.5   Cluster Representative Individual Selection and Variable K-Means Results

Several problems, during initial test using the KMS-based methodology for yield optimization, were identified and already discussed in Sect. 4.3. To overcome the false POF problem and the effects caused by the projection of the potential solutions yield value into the cluster representative individual yield line, several methods were presented [9]. Among those methods are different cluster representative individual

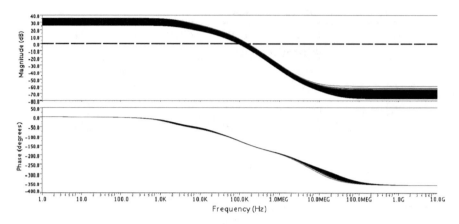

Fig. 6.14 OPDK single-stage amplifier Monte Carlo AC response

selection techniques cluster representative individual selection techniques and a variable number of clusters for the KMS-based methodology. The cluster representative individual selection presented in Sect. 4.3.4.1 include a method where the representative individual for each cluster is the potential solution with the largest distance to the boundaries of the feasibility region defined by the problem constraints conditions. Another method for the cluster representative individual selection combines the original selection method, i.e., the best potential solutions per cluster in terms of the objective(s) value(s), with the method where the potential solutions with the largest distance to the boundaries of the feasibility region are selected. The variable number of clusters during the optimization process methods, discussed in Sect. 4.3.4.2, are studied to assess how the effect of underestimating the yield for solutions close to the higher yield POF regions, caused by the projection of the yield values into the cluster representative individual yield line, may be avoided without increasing the number of clusters.

The test circuit adopted for this section is the silicon-based 130 nm technology node single-stage amplifier with enhanced DC gain circuit from Sect. 6.1.1. The optimization problem performance and functional constrains are detailed in Table 6.1; also the design variables and ranges are presented in Table 6.2. The optimization goals were the maximization of the yield and FoM given by (6.3).

The number of clusters for the new cluster representative individual selection methods was fixed to 10 and both variable cluster selection methods had a maximum of 10 clusters. The NSGA-II population was set to 256 individuals and the optimization process runs for 300 generations. The number of MC iterations per simulations was set to 500. The tests were carried on virtual machine environment running on an i7-3770 Intel© CPU with 16 GB of RAM, which allows performing multiple and different optimization runs at the same time. At the end of the tests the best approach was selected to run on the standalone server, in order to assess execution times in a regular working environment.

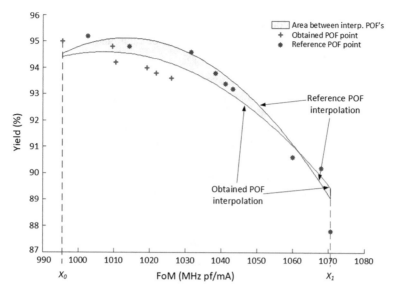

Fig. 6.15 Distance between two interpolated POFs

In order to assess the quality of the results and compare the tested methodologies, a measure that quantifies the distance between the obtained Pareto front by each methodology and the reference Pareto where all solutions were submitted to the MC simulations was developed. The new measure developed for an optimization problem where both objectives are being optimized is the average distance between the 2 s order polynomial interpolation curves defined by the points in each POF. This measure corresponds to the difference of areas defined by the interpolation curve in the interval between the minimum and maximum x-values of the interpolated points of both POFs; this idea is illustrated in Fig. 6.15, which is then divided by the integration range (6.7).

$$d_{av} = \frac{\int_{x_0}^{x_1} (p_1(x) - p_2(x))dx}{x_1 - x_0} \tag{6.7}$$

where

p_1—second-order polynomial interpolation for the reference POF
p_2—second-order polynomial interpolation for the obtained POF
x_0—minimum x value between the optimal and obtained POF points
x_1—maximum x value between the optimal and obtained POF points

In case (6.7) becomes negative ($d_{av} < 0$), the obtained and reference POF will be plotted in order to verify if the obtain solution dominates the reference POF, which only happens if the problem of the false POF, identified in Fig. 4.23, persists. The total MC simulations reduction percentage by each method was also registered, which works as an indicator of the total runtime of each methodology.

The five tested methods are identified as:

- *Best objective*: Cluster representative individual selection technique implemented at the initial KMS-based yield estimation and optimization method. This method selects as cluster representative individual the potential solution in each cluster with best optimization problem objective(s) value(s).
- *Large distance*: This cluster representative individual selection technique selects from each cluster the feasible potential solutions with the largest distance to the feasibility boundaries. The idea behind this technique is that potential solutions well inside the feasibility region present higher yield values.
- *Best objective + Large distance (Best obj. + L. dist.)*: As the name suggests, this technique combines the two previously presented techniques. The optimization problem objective(s), excluding the yield, and the distance to the feasibility boundaries are combined to create a multi-objective space where the selected cluster representative individual is the potential solutions closest to an ideal point, Fig. 4.32.
- *Exponential cluster decay (Exp. cluster decay)*: This method implements a technique where the optimization process starts with the maximum cluster number set and according to the total number of generations defined in the optimization algorithm the number of cluster decay in an exponential manner, Fig. 4.33a.
- *Exponential cluster growth (Exp. cluster growth)*: This method implements a technique where the optimization process starts with two clusters and exponentially evolve to the maximum number of clusters according to the generations of the optimization process, Fig. 4.33b.

For every tested methodology five runs were performed, except for *Large distance* method, in which after three runs it became clear that it was unable to correct the false POF problem, Fig. 6.16.

In Table 6.18 the results for each methodology are presented. The *Avg. MC reduction* column is the reduction rate computed by (6.1), using the average potential solutions that performed MC simulations with respect to the Full MC method.

Excluding the *Large distance* method, which was already identified as unable to solve the false POF problem, the other method with negative average distance values is the *Exponential decay* that may indicate the existence of the false POF problem, which after plotting the POFs was not verified, Fig. 6.17.

The results show that the variable exponential cluster number decay presents the best results considering the total number of MC simulations reduction rate and, also, the average distance to the ideal POF is quite close to the best objective approach. Based on the results, the exponential cluster number decay method was chosen to perform the final circuit sizing and optimization on the standalone working server in order to assess execution times.

The optimization process for the best approach reached the solutions presented in Fig. 6.18. The optimization process run for 300 generations, and took almost 41 min, which is an increase in speed performance of 50%, when comparing to the basic KMS approach.

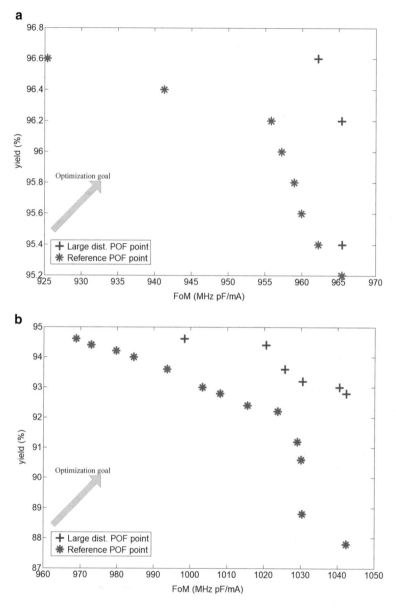

Fig. 6.16 Two runs for the *Large distance* method, showing false POF problem. (**a**) First run. False POF problem with only 3 solutions in the Pareto front. (**b**) Second run. Presents false POF problem with 6 solutions in the Pareto front

The obtained POF shows a set of solutions that reached a maximum yield of 95.4% with a FoM of 996 MHz pF/mA and a minimum yield of 89% for a FoM of 1082 MHz pF/mA. The final solution presents a POF with fewer points than the basic KMS work, but with higher values both in yield and FoM.

Table 6.18 Average results for the tested methodologies

Approach	Average distance to ideal POF			Avg. full MC simulations	Avg. method MC simulations	Avg. MC reduction (%)
	Min.	Avg.	Max.			
Best objective	−0.0164	0.0307	0.0774	4465	1130	74.7
Large distance	−4.393	−1.3263	1.7405	4382	1061	75.8
Best Obj. + L. Dist.	−0.3898	0.1805	0.6577	3954	969	75.5
Exp. cluster decay	0.0424	0.0974	0.1277	6442	549	91.5
Exp. cluster growth	−0.2073	−0.0238	0.0511	5891	1643	72.1

The new variable cluster number yield optimization process takes the same time as the four-corners optimization process for the same circuit-sizing problem, and with the advantage of offering a more suitable approach to optimize and obtain more robust solutions for the analog circuit-sizing problem.

The good results from the exponential decay number of clusters method are related to the correct setting for the number of clusters along the optimization process. At the beginning of the optimization process a small number of feasible potential solutions per generation exists, the number of potential solutions was less than the maximum number of clusters, so the few feasible potential solutions were all simulated providing accurate yield estimation. Near the end of the optimization process more feasible potential solutions are created in each generation, the optimization algorithm at this stage is in the exploitation phase where solutions are closer to each other making appropriate setting to two the number of clusters, which is enough to correctly estimate the yield for the rest of potential solutions in the clusters.

6.6 FUZYE Methodology Results

The main goal of the tests presented in this section was to assess the new FUZYE methodology yield estimation accuracy; also, the number of clusters reduction technique implemented is validated. In order to achieve this goal, the proposed FUZYE methodology and the two-phase evaluation process replaces the standard evaluation process in the AIDA-C analog IC sizing tool. The new implemented methodology in the AIDA-C evaluation process is compared with the previous KMS-based approach [4], and to the Full MC reference situation, where all potential solutions are subject to MC simulations for yield estimation. Based on this last yield value, from Full MC, it is possible to define a yield estimation error measure using (6.2) to assess the impact of using clustering algorithms to reduce the number of MC

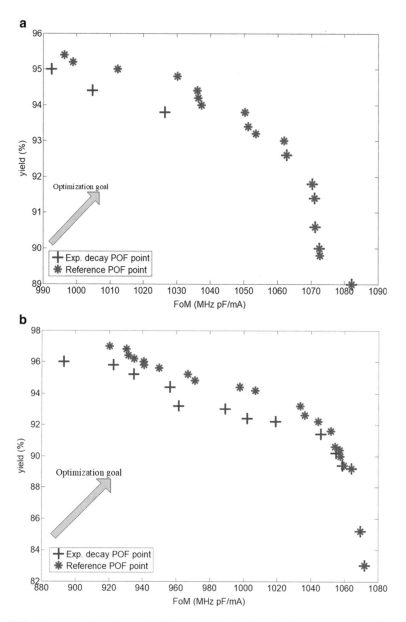

Fig. 6.17 Two runs of the *Exponential decay* method, showing that the problem of the false POF was solved

simulations. Additionally, tests using different fuzziness parameters values, m, are performed to identify the best value for this parameter.

Two circuit examples, which represent typical circuitry blocks in the analog world, were used to test the FUZYE methodology. The first circuit is the single-stage

Fig. 6.18 Final Pareto front for the variable cluster yield optimization process

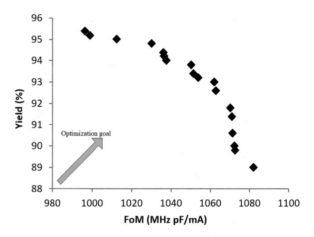

amplifier with enhanced DC gain from Sect. 6.1.1. The second circuit is the LNA for 5 GHz applications from Sect. 6.1.3. The adopted sizing and optimization process for both circuits perform MC simulations using the foundry PDK's statistical device models, considering process and mismatch variations. For both examples a 130 nm silicon technology node was used.

The GA population was set to 256 elements and as stopping criteria a maximum of 300 generations for the single-stage amplifier and 600 generations for the LNA were adopted. Additionally, Mentor's Eldo© circuit simulator was selected to calculate the circuit performance measures and to perform the MC simulations. The number of potential solutions simulated for MC analysis by each approach, i.e., KMS-based and FUZYE, were registered allowing the calculation of the number of saved or reduced simulations, when comparing to Full MC.

The number of MC iterations for each feasible potential solution was set to 1000. The relation between the number of MC iterations and the accuracy of the estimated yield can be expressed by (3.8), which in terms of multiples of the standard deviation of the distribution is given by (6.8).

$$k_y = \sqrt{\frac{N\, \Delta Y^2}{Y(1-Y)}} \qquad (6.8)$$

Using (6.8) it is possible to find the corresponding confidence level according to the yield estimated value. Setting the number of MC iterations $N = 1000$ and ΔY to 2.5% in (6.8), it is possible to verify that the estimated results confidence level is above 90% for yield values higher than 64%, as depicted in Fig. 6.19. Since the yield estimated values in the FUZYE methodology tests are higher than 80%, the obtained results present a confidence level above 95%.

The number of clusters considered for the KMS-based approach was set to five clusters. While, for the FUZYE methodology, the number of clusters was

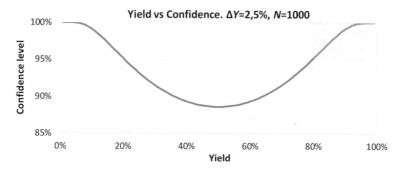

Fig. 6.19 Yield vs. confidence level for a 2.5% confidence interval and 1000 MC iterations

automatically set by the algorithm. Moreover, for the FUZYE methodology a range of different m parameter values were tested. The tests were carried on a server with an i7-3770 Intel© CPU with 16 GB of RAM, and, for each test, five runs were executed.

6.6.1   FUZYE Number of Clusters Reduction Validation

In this section, the cluster number reduction technique implemented in the FUZYE methodology, which is a feature that allows saving additional MC simulations during the optimization process, is validated.

The adopted cluster number reduction technique explores a necessary step, before the cluster representative individual selection, where all the elements must be allocated to their main clusters, FUZYE-Step 3 detailed in Fig. 4.41. In this step, the degree of membership matrix is examined and based on highest membership value each element is allocated to a specific cluster. Finished the allocation process, the clusters are examined, and if a cluster does not have any elements assigned, then the clustering process is repeated with minus one cluster than in the previous clustering iteration.

Assuming the correct choice for the number of clusters, by using Algorithm 4.6, the cluster reduction number technique will be validated using two cluster validation indexes. The selected validation indexes were already presented at Sect. 4.2.4, which are the partition coefficient (PC) (4.21) and partition entropy (PE) (4.22) indexes. Many other cluster validation indexes exist, but for the FUZYE cluster number reduction validation process the selected indexes are ideal, since each element is allocated to their main cluster using this membership value and, also, the adopted indexes are based on the degree of membership, which gives a measure of how well the clusters are defined.

The partition coefficient index measures the overlap between clusters, where for an index value of one no cluster overlap exists. So, for index values closer to one, better cluster separation is achieved. The partition entropy index also measures

Table 6.19   Validation for cluster number reduction technique

Index	Analysis for cluster number reduction decision				
	Avg. value after reduction	Avg. value with no reduction	Correct	Wrong	% Correct decision
Partition coefficient	0.38	0.24	110	3	97
Partition entropy	1.04	1.51	112	1	99

cluster overlapping, where the closer the index value is to zero, better defined clusters are obtained.

The circuit adopted to perform the validation test and compute the clusters validation indexes is the single-stage amplifier from Sect. 6.1.1. The fuzziness parameter of the FUZYE methodology was set to $m = 2$, and the optimization process run for 300 generations. At the MC netlist and test bench file the number of iterations per simulated solution was set to 500. The validation process was performed in 291 clustering runs, since in 9 of the 300 generations no feasible solutions were found. During the optimization process, the FUZYE methodology reduced the number of clusters in 113 occasions, and for 178 runs executed the clustering process with the initial number of clusters set by the Algorithm 4.6.

In order to assess if the reduction in clusters was the right decision, the partition coefficient index value after reducing the number of clusters must be higher than the index value obtained for the selected cluster number using Algorithm 4.6. Whereas for the partition entropy index, the index value must be lower when the number of clusters is reduced, since the objective goal is minimizing the partition entropy index.

The results in Table 6.19 show better index values for both indexes after the decision of reducing the clusters when empty clusters are found. In 113 cluster reductions, the algorithm took 110 correct decisions according to the partition coefficient index values and for the partition entropy index the decision was correct in 112 occasions. These values indicate that in 97% and 99% of the times reducing the number of clusters resulted in a more effective cluster separation, according to the validation indexes PC and PE, respectively.

The implemented technique, to reduce the number of clusters, resulted in a better cluster separation and in an additional reduction in the number of MC simulations performed by the FUZYE methodology.

6.6.2   FUZYE Single-Stage Amplifier Sizing and Optimization Results

The single-stage amplifier with enhanced DC gain performance specifications and functional constraints are detailed in Tables 6.1 and 6.2 and the optimization

problem design variables are presented. Like in previous examples, the maximization of the FoM and yield of the circuit are the optimization goals.

The yield estimation error measure (6.2) results comparing KMS-based and
FUZYE algorithm are presented in Table 6.20. The *improvement* columns values
in the several result tables presented are computed by (6.9).

$$\text{Improvement} = \frac{\text{KMS}_{\text{Est.Error/Simul.}} - \text{FUZYE}_{\text{Est.Error/Simul.}}}{\text{KMS}_{\text{Est.Error/Simul.}}} (\%) \qquad (6.9)$$

The results in Table 6.20 show that the FUZYE methodology attained lower yield
estimation error than the KMS-based methodology. The presented values were
obtained by setting $m = 2$. In Table 6.21, the average yield estimation error values
for an increasing value of the fuzziness parameter are presented. The yield estimation
error results show that the best value for the m parameter is 2, since it presents the
best yield estimation error among the different tested values.

In Table 6.22, the number of simulated potential solutions for MC analysis, the
reduction rate (6.1) when compared to Full MC, and the improvement of the FUZYE
methodology when compared with KMS are presented. The MC simulations reduction rate shows values above 80% for both approaches, but when comparing the
number of simulations among the KMS-based and FUZYE, the results revealed an
average improvement of 23% for the FUZYE methodology.

In Fig. 6.20, the POF for the different approaches in one of the optimization
runs is presented. This image shows that the FUZYE methodology provided
optimal solutions with the same values of the Full MC, while the KMS-based
approach presents two points that are dominated by the Full MC and the FUZYE
methodology.

The decision of clustering, in the design variable space instead of the specification
performance space, was based on 5 runs where both clustering spaces were tested.
After 300 generations the POF presented on average only four solutions for the
performance space cluster. Also, the solutions were dominated by the design variable space clustering approach, as illustrated in Fig. 6.21. The great similitude of
performance values among potential solutions at each generation results in a small
number of clusters, where the cluster representative individual clearly dominates the
rest of the potential solutions. This causes the small final POF for the performance
space clustering.

Table 6.20 FUZYE single-stage amplifier yield estimation error test results

Run	KMS-based est. error	FUZYE est. error ($m = 2$)	FUZYE improvement (%)
1	3.40e−2	3.08e−2	9.41
2	4.25e−2	3.80e−2	10.59
3	2.37e−2	2.07e−2	12.66
4	3.92e−2	3.47e−2	11.48
5	3.56e−2	2.83e−2	20.51
Average	*3.50e−2*	*3.05e−2*	*12.86*

Table 6.21 Average results for different m fuzziness parameter values in the FUZYE methodology

m fuzziness parameter	KMS-based est. error	FUZYE est. error	FUZYE improvement (%)
$m = 2$	3.50e−2	3.05e−2	12.86
$m = 3$	3.63e−2	3.87e−2	−6.61
$m = 4$	3.51e−2	4.12e−2	−17.38

Table 6.22 Single-stage amplifier results for the MC simulations reduction rate

	Full MC	K-means		FUZYE ($m = 2$)		
Run	Simul.	Simul.	Red. (%)	Simul.	Red. (%)	Improv. (%)
1	6687	1252	81	957	86	24
2	9482	1238	87	958	90	23
3	7632	1129	85	808	89	28
4	9831	1307	87	1017	90	22
5	8548	1264	85	1055	88	17
Average	*8436*	*1238*	*85*	*959*	*89*	*23*

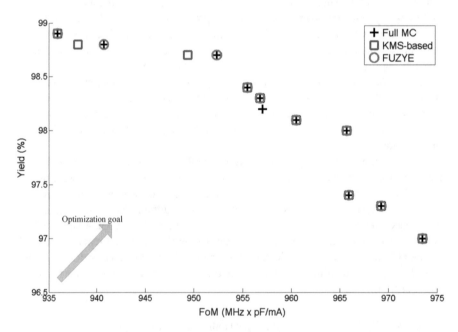

Fig. 6.20 Final POF achieved by the different approaches: Full MC, KMS-based, and FUZYE

Finally, two new optimization processes were performed using the single-stage amplifier, where the FoM, DC gain (GDC), and yield were maximized using only the FUZYE methodology.

The first optimization run implements the MC-based FUZYE methodology (MC-FUZYE) discussed so far, using pseudo-random sampling to generate the data to perform MC simulations. This new run achieved solutions with 99.8% of yield. In Fig. 6.22 an interpolated surface from the Pareto solutions is presented.

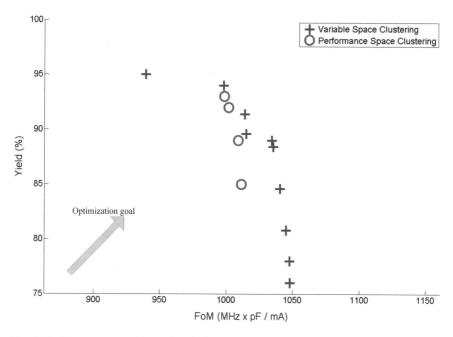

Fig. 6.21 Clustering potential solutions in the performance space vs. design variable space

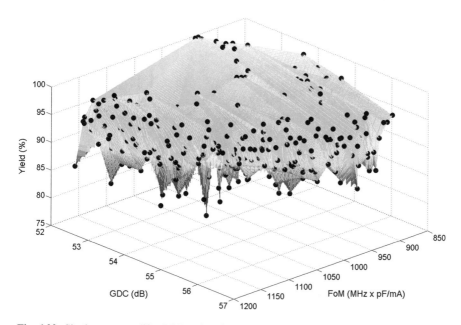

Fig. 6.22 Single-stage amplifier POF surface for three-objectives MC-FUZYE sizing and optimization run (FoM, GDC, Yield)

The MC-FUZYE three-objective optimization results were compared with a previous optimization where only the FoM and GDC were maximized and no yield optimization was performed. This comparison is shown in Fig. 6.23, where the MC-FUZYE solutions, which are the projection of the solutions in Fig. 6.22 into the FoM/GDC plane, are represented in different markers according to the yield interval they belong. As expected, the higher yield solutions are on the region most distant from the two-objective Pareto front, which presents solutions closer to feasibility region boundaries.

The MC-FUZYE three-objective run performed MC simulations in 839 potential solutions, achieving a simulation reduction rate of 91% when compared to Full MC optimization. The full optimization process was concluded in 1 h and 37 min. To validate the results of the three higher yield solutions (99.8%), identified by numbers in Fig. 6.23, 10 runs with 50,000 MC iterations were performed per solution. In Table 6.23, the average results for all the simulations runs are presented. These results include all the performance specification and its respective process capability index, the average yield value estimated by the 10 runs of 50,000 MC iterations, the estimated yield standard deviation, the confidence interval with a confidence level of 99% and the absolute yield error between the MC-FUZYE yield result and average yield for the 50,000 MC iterations.

The second single-stage amplifier optimization run adopts the QMC sampling, one of the available sampling techniques in the adopted circuit simulator. No modifications are needed in the FUZYE methodology when using QMC or other sampling techniques provided by simulator, only the circuit netlist files require some

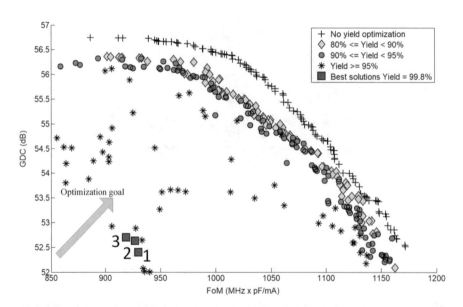

Fig. 6.23 Comparison between non-yield optimization process results and the FUZYE methodology

Table 6.23 FUZYE single-stage amplifier average results for the 10 runs of 50,000 MC iterations on each of the best yield solutions found

Sol.	I_{DD} (μA) ≤350		Gain (dB) ≥50		GBW (MHz) ≥30		PM (°) ≥60		FoM (MHz × pF/ mA) ≥850		Yield (%)	Std. dev. yield—σ	Confidence interval (99%)	Error (%)
	Avg.	C_{pk}	Avg.	C_{pk}	Avg.	C_{pk}	Avg.	C_{pk}	Avg.	C_{pk}				
1	339.76	0.23	51.9	0.66	52.51	3.89	67.3	4.50	928	2.97	99.62	0.0367	$99.52 \leq Y \leq 99.71$	0.184
2	335.30	0.33	52.1	0.67	51.63	3.77	64.8	2.78	924	2.78	99.66	0.0279	$99.59 \leq Y \leq 99.74$	0.136
3	332.16	0.41	52.1	0.67	50.73	3.61	67.6	4.58	917	2.62	99.68	0.0275	$99.61 \leq Y \leq 99.76$	0.115

minor changes to perform QMC-FUZYE. The adoption of QMC allows reducing the number of iterations performed on each PS. According to [10], in QMC satisfactory yield estimation results are obtained by considering iteration values between 25 and 50. For more accurate results, the number of iterations should increase to the range of hundreds; therefore, for the current test the number of iterations was set to 500. Also, in this QMC-FUZYE optimization run environment parameters variations were modelled. For that, the supply voltage variable was modelled as a Gaussian distribution with mean $\mu = 3.3$ V and standard deviation $\sigma = 5\%$ of the mean. Since the data is sampled in the interval $[\mu - 3\sigma, \mu + 3\sigma]$, the power supply variations may go from 2.8 to 3.8 V. The temperature operating conditions were modelled as a uniform distribution based on the temperature interval from $15°$ to $40°$. The stopping criterion in the QMC-FUZYE run was set to 300 generations, which were concluded in 1 h and 3 min. The QMC approach was able to reduce the execution time in 35% when compared to the MC-FUZYE approach. Although the number of iterations per potential solution used in the QMC approach was half of the MC approach, the execution time was not reduced proportional, mainly due to the overhead of invoking the circuit simulator. Considering the environment variable in the optimization processes resulted in solutions with lower yield values. The maximum yield solution achieved after the 300 generations was 88.2%. Clearly, in order to obtain higher yield values, to compensate the variability environment effects, the optimization process must continue beyond the 300 generations. So, the QMC-FUZYE optimization run continues up to 500 generations where higher yield values were achieved. This extended run was able to achieve a solution with 91.8% of yield. In Fig. 6.24,

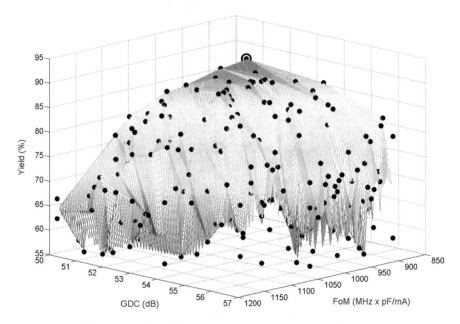

Fig. 6.24 Single-stage amplifier POF surface for three-objectives QMC-FUZYE sizing and optimization run with environment variability conditions

the obtained solutions are shown; also, the higher yield solution is presented surrounded by a circle. In order to validate the QMC sampling results, the higher yield value solution was tested with 50,000 MC iterations, which confirm the obtained value of 91.8% yield by the 500 QMC iterations.

6.6.3 FUZYE Low-Noise Amplifier for 5 GHz Applications Results

The second sizing and optimization circuit example, the LNA for 5 GHz applications, has its performance specification and functional constraints detailed in Table 6.6, and the optimization problem design variables and respective ranges at Table 6.7 are presented. The goals for this circuit are the maximization of the circuit parametric yield, the forward gain at 5.05 GHz (S_{21}) and the minimization of the LNA current consumption (I_{DD}). To calculate the circuit performance values, AC, noise, and steady-state analysis were performed by the electrical simulator.

The LNA circuit sizing and optimization process run for 600 generations for each of the 5 runs, achieving the yield estimation error results in Table 6.24.

The yield estimation error results show that using the FUZYE methodology a more accurate yield estimation is achieved, with an average improvement above 10% when compared to KMS. The results were obtained by setting the fuzziness parameter to $m = 2$. As in the previous circuit example, Sect. 6.6.2, several fuzziness parameter values were tested. In Table 6.25, the results for the different fuzziness parameter values are presented. Like for the previous circuit example, the smaller error was obtained for $m = 2$.

In Table 6.26, the MC simulations reduction rate with the KMS approach is 10%, while for the FUZYE method is almost 40%. Although this circuit performed

Table 6.24 FUZYE LNA yield estimation error test results

Run	K-means est. error	FUZYE est. error ($m = 2$)	FUZYE improvement (%)
1	1.91e−2	1.67e−2	12.57
2	4.09e−3	3.90e−3	4.65
3	7.37e−3	6.58e−3	10.72
4	9.77e−3	8.46e−3	13.41
5	1.60e−2	1.48e−2	7.44
Average	*1.13e−2*	*1.01e−2*	*10.44*

Table 6.25 Average results for different m fuzziness parameter values in the FUZYE LNA circuit

m fuzziness parameter	KMS-based est. error	FUZYE est. error	FUZYE improvement (%)
$m = 2$	1.13e−2	1.01e−2	10.44
$m = 3$	9.04e−3	9.41e−3	−4.09
$m = 4$	8.00e−3	9.35e−3	−16.88

Table 6.26 MC simulation reduction rate for the LNA circuit

	Full MC	K-means		FUZYE ($m = 2$)		
Run	Simul.	Simul.	Red. (%)	Simul.	Red. (%)	Improv. (%)
1	1902	1716	9.8	1133	40.4	33.9
2	1599	1411	11.8	936	41.5	33.6
3	1582	1477	6.6	1048	33.8	29.0
4	2242	1968	12.3	1273	432.	35.2
5	1812	1648	9.0	1122	38.1	32.9
Average	*1827*	*1644*	*10*	*1102*	*39.7*	*32.9*

600 generations, the number of potential solutions tested in MC is much smaller than expected. This fact is explained by the small number of feasible potential solutions at each generation, which also explains the smaller yield estimation error when compared to the previous tested circuit.

A new run with QMC for the LNA circuit was compared with a previous optimized circuit where no yield optimization was performed and only the forward gain (S_{21}) and current consumption (I_{DD}) were optimized. The QMC yield optimization run performed 500 QMC iterations at each feasible potential solution. The stopping criteria was set as 2250 generations, which were achieved after 55 h and 42 min for the two-objective optimization run with no yield optimization, and, 61 h and 49 min for the QMC-FUZYE methodology. The long optimization time was caused by the more complex analysis performed by the circuit simulator to extract the circuit performance measures. Also, to calculate the different measures two test bench files were used, implying multiple simulator calls per evaluation.

The difference of only 6 h between the two optimization runs, without considering yield and QMC-FUZYE, is explained by the small number of feasible potential solutions at each generation. In Fig. 6.25 the obtained POF surface is presented. The circles surrounding the solutions identify the top yield solutions, with values of 100% for the 500 iterations using QMC. Since the obtained yield value of 100% for 500 QMC iterations has some error associated, which is not possible to estimate when QMC is used [11], all points with this top yield value were validated with 50,000 MC iterations. This validation test shows yield values of 95.8%, 98.3%, 99.9%, and 99.8%, which correspond to the 100% points in Fig. 6.25, order from the lowest to the highest I_{DD} value. The results show an average error of 1.55% in the yield estimation using QMC with 500 iterations.

A comparison between the non-yield two-objective optimization and the QMC-FUZYE methodology is shown in Fig. 6.26. The QMC-FUZYE results are labelled according to the yield ranges they belong and were obtained by the projection of the three-objectives POF from Fig. 6.25 into the plane S_{21}/I_{DD}. As expected, potential solutions with highest yield values present smaller objective values than solutions in the non-yield-aware POF solutions.

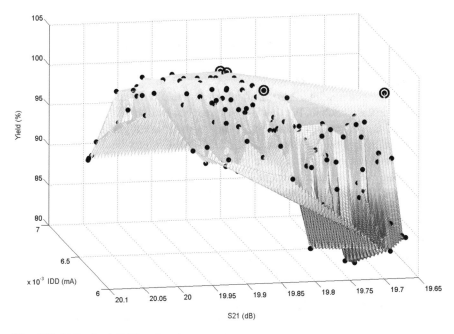

Fig. 6.25 LNA circuit POF surface for three-objectives QMC sampling sizing and optimization run after 2250 generations

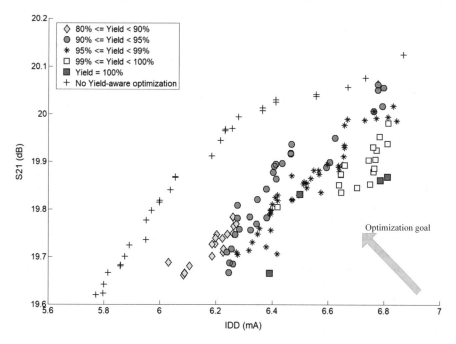

Fig. 6.26 LNA sizing and optimization POF solutions comparison between non-yield optimization and the QMC-FUZYE methodology

6.7   Many-Objective Optimization: A Comparison Between a Non-yield Approach and the FUZYE Methodology

The final test compares the old non-yield sizing and optimization process from AIDA-C with the new FUZYE methodology in a many-objective analog IC sizing and optimization problem. For this test was selected the filter circuit from Sect. 6.1.4 using a 65 nm technology node. The optimization problem performance specification and functional constraints are detailed in Table 6.8, and the 30 problem design variables and respective ranges are presented in Table 6.9. The MC simulations adopted the electrical simulator QMC sampling technique to perform the 500 iterations on each potential solution, and process and mismatch were considered on those simulations. The non-yield optimization process was set to run for 5000 generations, which took 83.5 h for completion. Based on the optimization time of the non-yield approach, the FUZYE methodology was manually stopped after the same amount of time. For that a large number of generations was set in order to assure that the optimization process did not stop before the desired time. The NSGA-II population was set to 128 individuals. The common optimization goals for the performance specification presented in Table 6.8 are the maximization of the *Gain DC*, *GBW*, and *PM*, and also the minimization of I_{DD}, *Offset*, and *Sdnoise*. Naturally for the FUZYE methodology an additional optimization goal was defined, which was the maximization of the yield.

The FUZYE methodology was able to find its first feasible solution at the 18th generation of the optimization algorithm, whereas the non-yield optimization found its first feasible solution at the 10th generation. Notice that before reaching a feasible solution the FUZYE methodology performs only the first evaluation phase as the non-yield optimization, so the difference of eight generation was due to the random initialization of the GA population.

The FUZYE methodology was able to perform 1606 generations in the 83.5 h. The total number of potential solutions that must be subject to MC simulations in a Full MC approach was 12,848 and FUZYE performed 5761 MC simulations, thus achieving a reduction rate of 55% according to (6.1).

In order to compare the results between both sizing and optimization processes, the final obtained Pareto solutions from each methodology were plotted. Since it is impossible to plot higher than three dimensions, in Fig. 6.27 the projections of the solutions into the objective's planes are presented. Graphics in Fig. 6.27 show each objective versus the *Gain DC* objective. The non-yield optimization methodology results are plotted as sum signs and the FUZYE results are represented as circles. The graphics also identify the best yield solution found by FUZYE, with a value of 85.8% of yield. Additionally are identified the best trade-off solutions from each methodology. The best trade-off solutions correspond to the solutions closer to the best ideal data point with component values that are the best objective values found by the optimization process, similarly with the idea depicted in Fig. 4.24. In Table 6.27 both considered ideal data points are detailed.

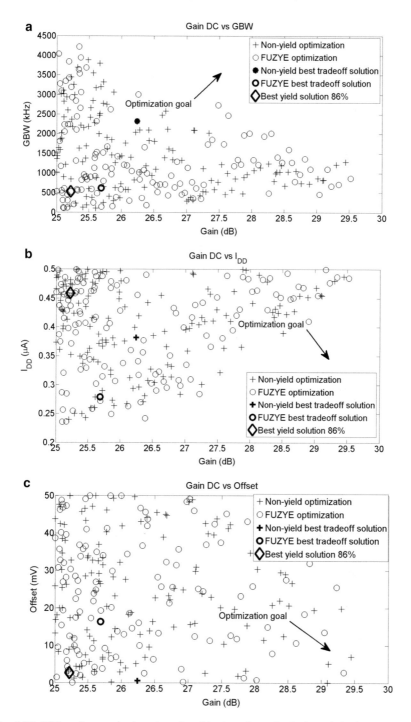

Fig. 6.27 POF surface projections into the objective planes for both optimization processes. (**a**) Gain DC vs GBW projection. (**b**) Gain DC vs I$_{DD}$ projection. (**c**) Gain DC vs Offset projection. (**d**) Gain DC vs PM projection. (**e**) Gain DC vs Sdnoise projection. (**f**) Gain DC vs Yield projection

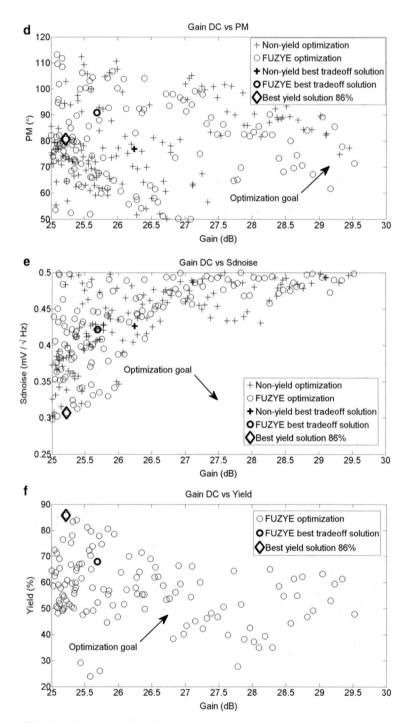

Fig. 6.27 (continued)

Table 6.27 Ideal data points for the non-yield and FUZYE methodologies

Methodology	Gain DC (dB)	GBW (kHz)	I_{DD} (µA)	Offset (mV)	PM (°)	Sdnoise (mV/$\sqrt{Hz}$)	Yield (%)
Non-yield	29.45	3878.1	0.239	0.0069	112	0.302	
FUZYE	29.52	4220.9	0.235	0.0030	113	0.298	85.8

Table 6.28 Best trade-off solutions for the non-yield, FUZYE and the best yield FUZYE solution

Best trade-off	Gain DC (dB)	GBW (kHz)	I_{DD} (µA)	Offset (mV)	PM (°)	Sdnoise (mV/$\sqrt{Hz}$)	Yield (%)
Non-yield	26.24	2323.4	0.381	0.549	76	0.426	42.6
FUZYE	25.69	619.2	0.279	16.29	90	0.421	68.0
Best yield	25.22	529.5	0.458	2.70	80	0.306	85.8

The best trade-off solutions attained by each methodology are detailed in Table 6.28; also the solution with best yield value is presented. The yield value presented for the best non-yield trade-off solutions was obtained by performing MC simulations using 500 iterations and QMC sampling.

The best trade-off solution of the non-yield optimization methodology presents a yield below 50%, which would force to a new optimization process considering larger safe margins for the circuit performance specifications. Although the best yield solution found by FUZYE methodology may present, in some of the performance specification, sub-optimal values, the fact is that considering the yield during the optimization process avoids performing new sizing and optimization processes with overdesign approaches.

6.8 Conclusions

In this chapter several different approaches to reduce the negative time impact of MC simulations in a population-based optimization loop were tested. The test results have shown that by using the new KMS-based methodology it was possible to achieve an average reduction in MC simulations of 75%. By further improving the KMS-based method with the variable cluster exponential decay technique, it was possible to achieve an average reduction of 91% in the total number of performed MC simulations needed to evaluate the full NSGA-II population, and furthermore a speedup of 50% of the optimization process in comparison with the simple KMS-based method.

In order to improve the trade-off between the number of MC simulations and yield estimation accuracy, the new FUZYE methodology was tested. The FUZYE methodology results show an average accuracy improvement of 13%, and a reduction rate improvement of 23% when compared to the KMS-based methodology. The FUZYE methodology achieved a high reduction rate in the total number of MC simulations, up to 81%, when compared to performing MC simulation to the full population of the optimization algorithm. The results validate the new FUZYE methodology showing that it is possible considering an expensive simulation technique, such as the MC simulations, to evaluate potential solutions inside the optimization loop of a population-based optimization algorithm. Moreover, considering the yield as an objective for analog IC sizing and optimization problems allows reducing the number of redesign iterations, since the robustness of solutions is being considered at early stages of the design flow.

References

1. R. Povoa, N. Lourenco, N. Horta, R. Santos-Tavares, J. Goes, Single-stage amplifiers with gain enhancement and improved energy-efficiency employing voltage-combiners, in *2013 IFIP/ IEEE 21st Int. Conf. Very Large Scale Integration (VLSI-SoC)*, 2013
2. M. Pandey, A. Canelas, R. Póvoa, J.A. Torres, J.C. Freire, N. Lourenço, N. Horta, Design and application of a CMOS active inductor at Ku band based on a multi-objective optimizer. Integration VLSI J **55**, 330–340 (2016)
3. A. Canelas, R. Póvoa, R. Martins, N. Lourenço, J. Guilherme, N. Horta, A 20 DB gain two-stage low-noise amplifier with high yield for 5 GHz applications, in *15th Int. Conf. Synthesis, Modeling, Anal. Simulation Methods Appl. Circuit Des. (SMACD)*, Prague, Czech Republic, 2018
4. A. Canelas, R. Martins, R. Póvoa, N. Lourenço, N. Horta, Yield optimization using k-means clustering algorithm to reduce Monte Carlo simulations, in *13th Int. Conf. Synthesis, Modeling, Anal. Simulation Methods Appl. Circuit Des. (SMACD)*, Lisbon, 2016
5. R. Spence, R. Soin, *Tolerance Design of Electronic Circuits* (Addison-Wesley, Wokingham, 1988)
6. M. Meehan, J. Purviance, *Yield and Reliability Design for Microwave Circuits and Systems* (Artech House, Norwood, MA, 1993)
7. K.Y. Mitra, E. Sowade, C. Martínez-Domingo, E. Ramon, J. Carrabina, H.L. Gomes, R.R. Baumann, Potential up-scaling of inkjet-printed devices for logical circuits in flexible electronics, in *Proc. AIP Conf.*, 2015
8. Fujifilm Dimatix, Materials Printer & Cartridge DMP-2800 Series Printer & DMC-11600 Series Cartridge FAQ Report, 21 Apr 2016
9. A. Canelas, R. Martins, R. Póvoa, N. Lourenço, N. Horta, Efficient Yield Optimization Method Using a Variable K-means Algorithm for Analog IC Sizing, in *2017 Des. Automat. Test Eur. Conf. Exhibition (DATE)*, Lausanne, 2017
10. M. Pak, F.V. Fernandez, G. Dundar, Comparison of QMC-based yield-aware pareto front techniques for multi-objective robust analog synthesis. Integration VLSI J. **55**, 357–365 (2016)
11. E. Afacan, G. Berkol, G. Dundar, A.E. Pusane, F. Baskaya, An analog circuit synthesis tool based on efficient and reliable yield estimation. Microelectron. J. **54**, 14–22 (2016)

Chapter 7
Conclusion and Future Work

7.1 Conclusions

Consumers' pressure for affordable electronic devices, offering new and greater features, brought new challenges for analog IC design. Additionally, the new nanometer-scale technology nodes adopted to produce today's electronic circuits revealed an increased sensitivity to variability effects, resulting in larger parametric yield loses.

In order to produce the demanded high-performance specifications ICs, a large number of performance and functional specification constraints must be enforced and satisfied, which led to the appearance of automatic analog IC sizing and optimization tools using heuristic optimization algorithms, such as AIDA-C [1], where potential solutions are accurately evaluated by electrical simulation. Those optimization algorithms are able to achieve optimal solutions in a reasonable amount of time by exploring the solution space with large number of potential solutions. To address yield loses, analog IC sizing and optimization tools using heuristic-based optimization algorithms must incorporate a methodology to estimate the yield of potential solutions and consider the yield as one of the optimization objectives of the sizing problem. The most accurate technique to estimate a circuit yield is by performing MC analysis based on electrical simulations using foundry PDK's statistical device models. Implementing such a solution to estimate the yield, i.e., MC analysis using electrical simulation, for all potential solutions of the optimization algorithm, may increase the overall optimization time to unacceptable values.

In this work, a methodology with reduced time impact to address the problem of analog IC yield estimation by means of MC analysis inside an optimization loop of a population-based algorithm was developed. The low time impact on the overall optimization processes offers IC designers the possibility of performing yield optimization with the most accurate yield estimation method, MC simulations using foundry statistical device models considering local and global variations.

© Springer Nature Switzerland AG 2020
A. M. L. Canelas et al., *Yield-Aware Analog IC Design and Optimization in Nanometer-scale Technologies*, https://doi.org/10.1007/978-3-030-41536-5_7

The new cluster-based yield estimation methodology selects from each cluster the potential solutions with the best objective values, which are most likely to become part of the POF, to perform the MC analysis and based on the simulated results estimates the yield for the rest of potential solutions. The initial approach to develop a new yield estimation methodology was based on a modified implementation of the KMS clustering algorithm. Results using the KMS-based approach showed a reduction rate up to 75% in the total number of MC simulations when compared to an optimization approach where all potential solutions are subject to MC analysis. A later implementation of KMS using a variable number of clusters during the optimization processes was able to achieve reduction rates up to 91%; the improvement in the reduction rate resulted in an increase in the yield estimation error.

In order to balance the trade-off between simulation reduction rate and yield estimation error, a new methodology based on the FCM algorithm was developed. The new methodology, named FUZYE, achieved a simulation reduction rate up to 81% with respect to the total number of required MC simulations for the complete population of the optimization algorithm. When compared to the KMS-based methodology, FUZYE was able to improve the simulation reduction rate by 23%, and with an increase in the yield estimation accuracy up to 13%. The improvement in yield estimation accuracy results from applying the degree of membership matrix to estimate the yield of non-simulated potential solutions based on the yield values of the cluster representative individuals, which were subject to MC analysis.

In Table 7.1, the new FUZYE methodology is compared with other state-of-the-art yield estimation and optimization approaches.

The new FUZYE methodology was successfully implemented in a state-of-the-art analog IC sizing and optimization tool known as AIDA-C, taking advantage of the user-friendly GUI already developed. The results show that the use of traditional "off-the-shelf" MC simulations using accurate foundry PDK's statistical models inside the optimization processes of AIDA was possible because only a small number and relevant solutions carefully selected by the new FUZYE methodology were subject to MC analysis. Additionally, IC designers are able to control all MC simulation parameters, like the number of iterations which allows adjusting the estimation accuracy, since simulations are based on a netlist and test bench file available for editing.

As the new FUZYE methodology relies on electrical simulators to perform MC simulations, all sampling techniques, such as QMC or LHS, provided by the electrical simulator can be adopted without any changes in the FUZYE methodology, which can further reduce the time impact of MC simulations. Along with the adoption of a sampling technique, the time impact of MC simulations can be reduced by parallelizing simulations, which is a feature now available on most commercial electrical simulators, if the number of available licenses so permits. To overcome the license limitation, AIDA development team is now working on a new release for AIDA-C using an open-source electrical simulator.

The developed FUZYE algorithm, which was tested for analog IC parametric yield estimation, achieved a high reduction in the number of MC simulations, and can also be used in other evaluation processes where expensive simulations are involved.

Table 7.1 Comparison between FUZYE methodology and other yield estimation approaches

	Monte Carlo based		Importance sampling		
	FUZYE	MC/Quasi-MC	Selective sampling	Model based	Non-Monte Carlo based
Works	This work	Singhee [2] Liu [3] Guerra-Gomez [4] Afacan [5–7] Pak [8]	Qazi [9] Yilmaz [10] Kanj [11] McConaghy [12] Yao [13] Wang [14] Sun [15] Kuo [16]	Kuo [16] Okobiah [17] Felt [18]	Li [19] Graeb [20] Bűrmen [21] Pan [22] Sciacca [23] Opalski [24]
Strengths	Accuracy. Performs MC on best non-dominated solutions to accurately estimate its yield. Can handle a large number of input process variables. If available by the electrical simulator can perform QMC or LHS with no changes in the algorithm.	Accuracy. Can handle a large number of input process variables.	High-sigma analysis. Explores failure region near the distribution tails.	Fast yield estimation	High-sigma analysis. Fast yield estimation.
Weaknesses	Time consuming for high sigma.	Time consuming.	Distorts the sampling distribution. Limited input process variables.	Accuracy. Reusability of the models.	Accuracy. Limited input process variables.
Strategy	Performs MC only on relevant potential solutions and for the rest of solutions accurately estimates yield based on tested solution.	QMC/LHS evenly sample parameter space. or Two-step MC analysis to identify promising solutions.	Samples the space around the failure region. or Selects regions of interest to perform sampling.	Uses models for fast MC simulations.	Performs at least one additional simulation per parameter variable to create a model for WC. Yield is indirectly estimated based on multiples of σ.

7.2   Future Work

Possible future areas of research are outlined in this section. Among those areas are topics not only related to improvements in the AIDA framework but also related to constraint optimization problems.

- *Development of a benchmark library of analog circuits.* One of the major difficulties validating the new yield estimation and optimization methodology was the lack of comparable results from other works in the area. This difficulty is present on all referenced works by this book, where the authors were only able to fairly compare their new results with their previous works. Most relevant works on analog IC sizing with yield optimization do not reveal all the required details, i.e., technology adopted, functional and performance constraints considered, the design variables and their ranges, to allow its reproduction and obtain results to compare. Also, some of the adopted technology nodes by other works are not available. So, to be able to fairly compare among different yield estimation methodologies and also analog IC sizing and optimization techniques, a generic and commonly set of analog circuits sizing problems must be created and freely distributed.
- *Automatic generation of the netlist and test bench file for MC analysis.* The new FUZYE methodology requires the definition of a netlist and test bench files to invoke the electrical simulator and perform the MC simulations. The MC netlist and test bench files are now manually created by the IC designer based on a copy of the typical simulation files. The automatic generation of the MC netlist and test bench files will automatically create the copy of the typical simulation files and add the set of commands to perform MC simulations according to the selected electrical simulator. Additionally, the statistical device models selected at the setup screen of AIDA-C will replace the typical models at the netlist files.
- *Creating a process to perform sensitivity analysis on selected solutions.* The new implemented feature of AIDA-C for yield optimization computes the process capability index for every performance considered relevant in terms of yield. Another important information for IC designers, which is also important during typical sizing and optimization processes, is the sensitivity analysis of the circuit performance specifications with respect to the design variables. This information reveals which design variable has more impact on performances, which permits designers to speed up the optimization processes.
- *New GA mutation operator.* Another line of research was detected during the circuit sizing and yield optimizations tests. The high number of constraints typically adopted by analog IC sizing problems creates small regions of feasibility, which results in a reduced number of feasible offspring solutions at each generation of the optimization algorithm, even at the exploitation phase of the algorithm. This means that at each generation a high number of simulations are spent evaluating infeasible individuals. An approach able to mutate infeasible offspring into the feasibility regions would represent an important efficiency improvement in terms of time and resources allocated to the circuit sizing and optimization processes. Based on this idea, a new approach is already under development, and encouraging results were already obtained.

References

1. N. Lourenço, R. Martins, N. Horta, *Automatic Analog IC Sizing and Optimization Constrained with PVT Corners and Layout Effects* (Springer International Publishing, Cham, 2017)
2. A. Singhee, R.A. Rutenbar, Why quasi-Monte Carlo is better than Monte Carlo or Latin hypercube sampling for statistical circuit analysis. IEEE Trans. Comput. Aided Des. Integr. Circuits Syst. **29**(11), 1763–1776 (2010)
3. B. Liu, F.V. Fernandez, G.G.E. Gielen, Efficient and accurate statistical analog yield optimization and variation-aware circuit sizing based on computational intelligence techniques. IEEE Trans. Comput. Aided Des. Integr. Circuits Syst. **30**(6), 793–805 (2011)
4. I. Guerra-Gomez, E. Tlelo-Cuautle, L.G. de la Fraga, OCBA in the yield optimization of analog integrated circuits by evolutionary algorithms, in *2015 IEEE Int. Symp. Circuits Syst. (ISCAS)*, Lisbon, 2015
5. E. Afacan, G. Berkol, A.E. Pusane, G. Dündar, F. Başkaya, Adaptive sized Quasi-Monte Carlo based yield aware analog circuit optimization tool, in *2014 5th Eur. Workshop on CMOS Variability (VARI)*, Palma de Mallorca, 2014
6. E. Afacan, G. Berkol, A.E. Pusane, G. Dündar, F. Başkaya, A hybrid Quasi Monte Carlo method for yield aware analog circuit sizing tool, in *2015 Des. Automat. Test Eur. Conf. Exhibition (DATE)*, Grenoble, 2015
7. E. Afacan, G. Berkol, G. Dundar, A.E. Pusane, F. Baskaya, An analog circuit synthesis tool based on efficient and reliable yield estimation. Microelectron. J. **54**, 14–22 (2016)
8. M. Pak, F.V. Fernandez, G. Dundar, Comparison of QMC-based yield-aware pareto front techniques for multi-objective robust analog synthesis. Integration VLSI J. **55**, 357–365 (2016)
9. M. Qazi, M. Tikekar, L. Dolecek, D. Shah, A. Chandrakasan, Loop flattening & spherical sampling: highly efficient model reduction techniques for SRAM yield analysis, in *2010 Des. Autom. Test Eur. Conf. Exhibition (DATE)*, Dresden, 2010
10. E. Yilmaz, S. Ozev, Adaptive-learning-based importance sampling for analog circuit DPPM estimation. IEEE Des. Test **32**(1), 36–43 (2015)
11. R. Kanj, R. Joshi and S. Nassif, Mixture importance sampling and its application to the analysis of SRAM designs in the presence of rare failure events, in *2006 43rd ACM/IEEE Design Automat. Conf.*, San Francisco, CA, 2006
12. T. McConaghy, K. Breen, J. Dyck, A. Gupta, *Variation-Aware Design of Custom Integrated Circuits: A Hands-on Field Guide* (Springer, New York, 2013)
13. J. Yao, Z. Ye, Y. Wang, Importance boundary sampling for SRAM yield analysis with multiple failure regions. IEEE Trans. Comput. Aided Des. Integr. Syst. **33**(3), 384–396 (2014)
14. M. Wang, C. Yan, X. Li, D. Zhou, X. Zeng, High-dimensional and multiple-failure-region importance sampling for SRAM yield analysis. IEEE Trans. Very Large Scale Integr. (VLSI) Syst. **25**(3), 806–819 (2017)
15. S. Sun, X. Li, H. Liu, K. Luo, B. Gu, Fast statistical analysis of rare circuit failure events via scaled-sigma sampling for high-dimensional variation space. IEEE Trans. Comput. Aided Des. Integr. Circuits Syst. **34**(7), 1096–1109 (2015)
16. C.C. Kuo, W.Y. Hu, Y.H. Chen, J.F. Kuan, Y.K. Cheng, Efficient trimmed-sample Monte Carlo methodology and yield-aware design flow for analog circuits, in *DAC Des. Autom. Conf.*, San Francisco, CA, 2012
17. O. Okobiah, S.P. Mohanty, E. Kougianos, Fast statistical process variation analysis using universal Kriging metamodeling: a PLL example, in *2013 IEEE 56th Int. Midwest Symp. Circuits Syst. (MWSCAS)*, Columbus, OH, 2013
18. E. Felt, S. Zanella, C. Guardiani, A. Sangiovanni-Vincentelli, Hierarchical statistical characterization of mixed-signal circuits using, in *Proc. Int. Conf. Comput. Aided Des.*, San Jose, CA, 1996
19. X. Li, W. Zhang, F. Wang, Large-scale statistical performance modeling of analog and mixed-signal circuits, in *Proc. IEEE 2012 Custom Integr. Circuits Conf.*, San Jose, CA, 2012
20. H.E. Graeb, *Analog Design Centering and Sizing* (Springer, Dordrecht, 2007)

21. Á. Bűrmen, H. Habal, Computing worst-case performance and yield of analog integrated circuits by means of mesh adaptive direct search. J. Microelectron. Electron. Comp. Mater. **45**(2), 160–170 (2015)
22. X. Pan, H. Graeb, Reliability analysis of analog circuits using quadratic lifetime worst-case distance prediction, in *IEEE Custom Integrated Circuits Conf.* 2010, San Jose, CA, 2010
23. E. Sciacca, S. Spinella, A.M. Anile, Possibilistic worst case distance and applications to circuit sizing, in *Theoretical Advances and Applications of Fuzzy Logic and Soft Computing*, (Springer, Berlin, 2007), pp. 287–295
24. L. Opalski, Remarks on statistical design centering. Int. J. Electron. Telecommun. **57**(2), 159–167 (2011)

Index

© Springer Nature Switzerland AG 2020
A. M. L. Canelas et al., *Yield-Aware Analog IC Design and Optimization in
Nanometer-scale Technologies*, https://doi.org/10.1007/978-3-030-41536-5

Printed in the United States
by Baker & Taylor Publisher Services